Also by Elizabeth Kolbert

Field Notes from a Catastrophe: Man, Nature, and Climate Change

The Prophet of Love: And Other Tales of Power and Deceit

THE SIXTH EXTINCTION

THE SIXTH EXTINCTION

AN UNNATURAL HISTORY

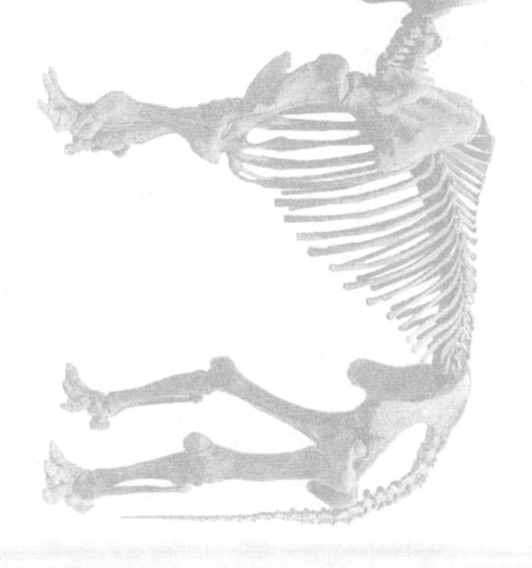

ELIZABETH KOLBERT

Henry Holt and Company New York

Henry Holt and Company, LLC
Publishers since 1866
175 Fifth Avenue
New York, New York 10010
www.henryholt.com

Distributed in Canada by Raincoast Book Distribution Limited

Library of Congress Cataloging-in-Publication Data

Kolbert, Elizabeth.
The sixth extinction : an unnatural history / Elizabeth Kolbert.
pages cm
Includes bibliographical references and index.
ISBN 978-0-8050-9299-8 (hardback)—ISBN 978-0-8050-9311-7 (electronic book) 1. Mass extinctions. 2. Extinction (Biology) 3. Environmental disasters. I. Title.
QE721.2.E97K65 2014
576.8′4—dc23

2013028683

Henry Holt books are available for special promotions and premiums. For details contact: Director, Special Markets.

First Edition 2014

Designed by Meryl Sussman Levavi

Printed in the United States of America

10 9 8 7 6 5 4 3 2 1

If there is danger in the human trajectory, it is not so much in the survival of our own species as in the fulfillment of the ultimate irony of organic evolution: that in the instant of achieving self-understanding through the mind of man, life has doomed its most beautiful creations.

—E. O. Wilson

Centuries of centuries and only in the present do things happen.

—Jorge Luis Borges

CONTENTS

AUTHOR'S NOTE

Though the discourse of science is metric, most Americans think in terms of miles, acres, and degrees Fahrenheit. All the figures in this book are given in English units, except where specially noted.

THE SIXTH EXTINCTION

PROLOGUE

Beginnings, it's said, are apt to be shadowy. So it is with this story, which starts with the emergence of a new species maybe two hundred thousand years ago. The species does not yet have a name—nothing does—but it has the capacity to name things.

As with any young species, this one's position is precarious. Its numbers are small, and its range restricted to a slice of eastern Africa. Slowly its population grows, but quite possibly then it contracts again—some would claim nearly fatally—to just a few thousand pairs.

The members of the species are not particularly swift or strong or fertile. They are, however, singularly resourceful. Gradually they push into regions with different climates, different predators, and different prey. None of the usual constraints of habitat or geography seem to check them. They cross rivers, plateaus, mountain ranges. In coastal regions, they gather shellfish; farther inland, they hunt mammals. Everywhere they settle, they adapt and innovate. On reaching Europe, they encounter creatures very much like themselves, but stockier and probably brawnier, who have been living on the continent far longer. They interbreed with these creatures and then, by one means or another, kill them off.

The end of this affair will turn out to be exemplary. As the species expands its range, it crosses paths with animals twice, ten, and even twenty times its size: huge cats, towering bears, turtles as big as elephants, sloths that stand fifteen feet tall. These species are more powerful and often fiercer. But they are slow to breed and are wiped out.

Although a land animal, our species—ever inventive—crosses the sea. It reaches islands inhabited by evolution's outliers: birds that lay foot-long eggs, pig-sized hippos, giant skinks. Accustomed to isolation, these creatures are ill-equipped to deal with the newcomers or their fellow travelers (mostly rats). Many of them, too, succumb.

The process continues, in fits and starts, for thousands of years, until the species, no longer so new, has spread to practically every corner of the globe. At this point, several things happen more or less at once that allow *Homo sapiens*, as it has come to call itself, to reproduce at an unprecedented rate. In a single century the population doubles; then it doubles again, and then again. Vast forests are razed. Humans do this deliberately, in order to feed themselves. Less deliberately, they shift organisms from one continent to another, reassembling the biosphere.

Meanwhile, an even stranger and more radical transformation is under way. Having discovered subterranean reserves of energy, humans begin to change the composition of the atmosphere. This, in turn, alters the climate and the chemistry of the oceans. Some plants and animals adjust by moving. They climb mountains and migrate toward the poles. But a great many—at first hundreds, then thousands, and finally perhaps millions—find themselves marooned. Extinction rates soar, and the texture of life changes.

No creature has ever altered life on the planet in this way

before, and yet other, comparable events have occurred. Very, very occasionally in the distant past, the planet has undergone change so wrenching that the diversity of life has plummeted. Five of these ancient events were catastrophic enough that they're put in their own category: the so-called Big Five. In what seems like a fantastic coincidence, but is probably no coincidence at all, the history of these events is recovered just as people come to realize that they are causing another one. When it is still too early to say whether it will reach the proportions of the Big Five, it becomes known as the Sixth Extinction.

The story of the Sixth Extinction, at least as I've chosen to tell it, comes in thirteen chapters. Each tracks a species that's in some way emblematic—the American mastodon, the great auk, an ammonite that disappeared at the end of the Cretaceous alongside the dinosaurs. The creatures in the early chapters are already gone, and this part of the book is mostly concerned with the great extinctions of the past and the twisting history of their discovery, starting with the work of the French naturalist Georges Cuvier. The second part of the book takes place very much in the present—in the increasingly fragmented Amazon rainforest, on a fast-warming slope in the Andes, on the outer reaches of the Great Barrier Reef. I chose to go to these particular places for the usual journalistic reasons—because there was a research station there or because someone invited me to tag along on an expedition. Such is the scope of the changes now taking place that I could have gone pretty much anywhere and, with the proper guidance, found signs of them. One chapter concerns a die-off happening more or less in my own backyard (and, quite possibly, in yours).

If extinction is a morbid topic, mass extinction is, well, massively so. It's also a fascinating one. In the pages that follow, I try to convey both sides: the excitement of what's being learned as well as the horror of it. My hope is that readers of this book will come away with an appreciation of the truly extraordinary moment in which we live.

CHAPTER I

THE SIXTH EXTINCTION

Atelopus zeteki

THE TOWN OF EL VALLE DE ANTÓN, IN CENTRAL PANAMA, SITS in the middle of a volcanic crater formed about a million years ago. The crater is almost four miles wide, but when the weather is clear you can see the jagged hills that surround the town like the walls of a ruined tower. El Valle has one main street, a police station, and an open-air market. In addition to the usual assortment of Panama hats and vividly colored embroidery, the market offers what must be the world's largest selection of golden-frog figurines. There are golden frogs resting on leaves and golden frogs sitting up on their haunches and—rather more difficult to understand—golden frogs clasping cell phones. There are golden frogs wearing frilly skirts and golden frogs striking dance poses and golden frogs smoking cigarettes through a holder, after the fashion of FDR. The golden frog, which is taxicab yellow with dark brown splotches, is endemic to the area around El Valle. It is considered a lucky symbol in Panama; its image is (or at least used to be) printed on lottery tickets.

As recently as a decade ago, golden frogs were easy to spot in the hills around El Valle. The frogs are toxic—it's been calculated that the poison contained in the skin of just one animal could kill a thousand average-sized mice—hence the vivid color, which makes them stand out against the forest floor. One creek not far from El Valle was nicknamed Thousand Frog Stream. A person walking along it would see so many golden frogs sunning themselves on the banks that, as one herpetologist who made the trip many times put it to me, "it was insane—absolutely insane."

Then the frogs around El Valle started to disappear. The problem—it was not yet perceived as a crisis—was first noticed to the west, near Panama's border with Costa Rica. An American graduate student happened to be studying frogs in the rainforest there. She went back to the States for a while to write her dissertation, and when she returned, she couldn't find any frogs or, for that matter, amphibians of any kind. She had no idea what was going on, but since she needed frogs for her research, she set up a new study site, farther east. At first the frogs at the new site seemed healthy; then the same thing happened: the amphibians vanished. The blight spread through the rainforest until, in 2002, the frogs in the hills and streams around the town of Santa Fe, about fifty miles west of El Valle, were effectively wiped out. In 2004, little corpses began showing up even closer to El Valle, around the town of El Copé. By this point, a group of biologists, some from Panama, others from the United States, had concluded that the golden frog was in grave danger. They decided to try to preserve a remnant population by removing a few dozen of each sex from the forest and raising them indoors. But whatever was killing the frogs was moving even faster than the biologists had feared. Before they could act on their plan, the wave hit.

I first read about the frogs of El Valle in a nature magazine for children that I picked up from my kids. The article, which was illustrated with full-color photos of the Panamanian golden frog and other brilliantly colored species, told the story of the spreading scourge and the biologists' efforts to get out in front of it. The biologists had hoped to have a new lab facility constructed in El Valle, but it was not ready in time. They raced to save as many animals as possible, even though they had nowhere to keep them. So what did they end up doing? They put them "in a frog hotel, of course!" The "incredible frog hotel"—really a local bed and breakfast—agreed to let the frogs stay (in their tanks) in a block of rented rooms.

"With biologists at their beck and call, the frogs enjoyed first-class accommodations that included maid and room service," the article noted. The frogs were also served delicious, fresh meals—"so fresh, in fact, the food could hop right off the plate."

Just a few weeks after I read about the "incredible frog hotel," I ran across another frog-related article written in a rather different key. This one, which appeared in the *Proceedings of the National Academy of Sciences*, was by a pair of herpetologists. It was titled "Are We in the Midst of the Sixth Mass Extinction? A View from the World of Amphibians." The authors, David Wake, of the University of California-Berkeley, and Vance Vredenburg, of San Francisco State, noted that there "have been five great mass extinctions during the history of life on this planet." These extinctions they described as events that led to "a profound loss of biodiversity." The first took place during the late Ordovician period, some 450 million years ago, when living things were still mainly confined to the water. The most devastating took place at the end of the Permian period, some 250 million years ago, and it came perilously close to emptying the earth out altogether. (This event is sometimes referred to as "the mother of mass extinctions" or "the great dying.") The most recent—and famous—mass extinction came at the close of the Cretaceous period; it wiped out, in addition to the dinosaurs, the plesiosaurs, the mosasaurs, the ammonites, and the pterosaurs. Wake and Vredenburg argued that, based on extinction rates among amphibians, an event of a similarly catastrophic nature was currently under way. Their article

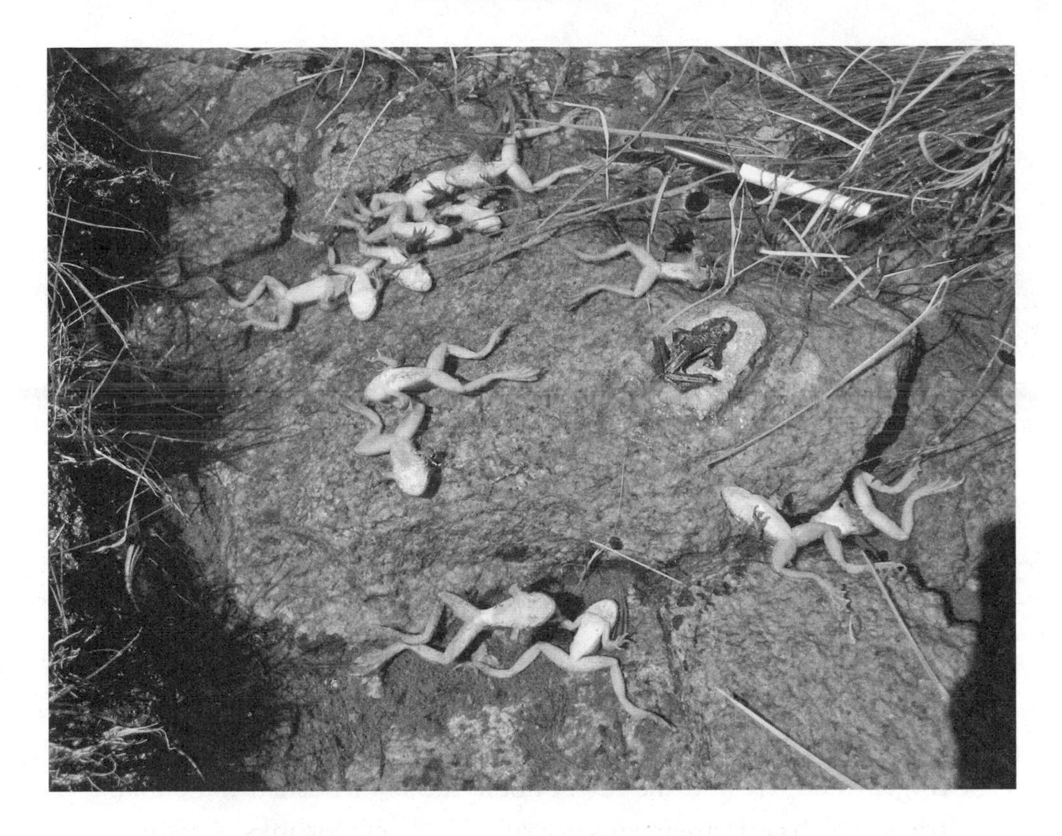

was illustrated with just one photograph, of about a dozen mountain yellow-legged frogs—all dead—lying bloated and belly-up on some rocks.

I understood why a kids' magazine had opted to publish photos of live frogs rather than dead ones. I also understood the impulse to play up the Beatrix Potter–like charms of amphibians ordering room service. Still, it seemed to me, as a journalist, that the magazine had buried the lede. Any event that has occurred just five times since the first animal with a backbone appeared, some five hundred million years ago, must qualify as exceedingly rare. The notion that a sixth such event would be taking place right now, more or less in front of our eyes, struck me as, to use the technical term, mind-boggling. Surely this story, too—the bigger, darker, far more consequential one—deserved telling. If Wake and Vredenburg were correct, then those of us alive today not only are witnessing one of the rarest events in life's history, we are also causing it. "One weedy species," the pair observed, "has unwittingly achieved the ability to directly affect its own fate and that of most of the other species on this planet." A few days after I read Wake and Vredenburg's article, I booked a ticket to Panama.

THE El Valle Amphibian Conservation Center, or EVACC (pronounced "ee-vac"), lies along a dirt road not far from the open-air market where the golden frog figurines are sold. It's about the size of a suburban ranch house, and it occupies the back corner of a small, sleepy zoo, just beyond a cage of very sleepy sloths. The entire building is filled with tanks. There are tanks lined up against the walls and more tanks stacked at the center of the room, like books on the shelves of a library. The taller tanks are occupied by species like the lemur tree frog, which lives in the forest canopy; the shorter tanks serve for species like the big-headed robber frog, which lives on the forest floor. Tanks of horned marsupial frogs, which carry their eggs in a pouch, sit next to tanks of casque-headed frogs, which carry their eggs on their backs. A few dozen tanks are devoted to Panamanian golden frogs, *Atelopus zeteki*.

Golden frogs have a distinctive, ambling gait that makes them look a bit like drunks trying to walk a straight line. They have long, skinny limbs, pointy yellow snouts, and very dark eyes, through which they seem to be regarding the world warily. At the risk of sounding weak-minded, I will say that they look intelligent. In the wild, females lay their eggs in shallow running water; males, meanwhile, defend their territory from the tops of mossy rocks. In EVACC, each golden frog tank has its own running water, provided by its own little hose, so that the animals can breed near a simulacrum of the streams that were once their home. In one of the ersatz streams, I noticed a strings of little pearl-like eggs. On a white board nearby someone had noted excitedly that one of the frogs *"depositó huevos!!"*

EVACC sits more or less in the middle of the golden frog's

A Panamanian golden frog (*Atelopus zeteki*).

range, but it is, by design, entirely cut off from the outside world. Nothing comes into the building that has not been thoroughly disinfected, including the frogs, which, in order to gain entry, must first be treated with a solution of bleach. Human visitors are required to wear special shoes and to leave behind any bags or knapsacks or equipment that they've used out in the field. All of the water that enters the tanks has been filtered and specially treated. The sealed-off nature of the place gives it the feel of a submarine or, perhaps more aptly, an ark mid-deluge.

EVACC's director is a Panamanian named Edgardo Griffith. Griffith is tall and broad-shouldered, with a round face and a wide smile. He wears a silver ring in each ear and has a large tattoo of a toad's skeleton on his left shin. Now in his mid-thirties, Griffith has devoted pretty much his entire adult life to the amphibians of El Valle, and he has turned his wife, an American who came to Panama as a Peace Corps volunteer, into a frog person, too. Griffith was the first person to notice when little carcasses started showing up in the area, and he personally collected many of the several hundred amphibians that got booked into the hotel. (The animals were transferred to EVACC once the building had been completed.) If EVACC is a sort of ark, Griffith becomes its Noah, though one on extended duty, since already he's been at things a good deal longer than forty days. Griffith told me that a key part of his job was getting to know the frogs as individuals. "Every one of them has the same value to me as an elephant," he said.

The first time I visited EVACC, Griffith pointed out to me the representatives of species that are now extinct in the wild. These included, in addition to the Panamanian golden frog, the Rabb's fringe-limbed tree frog, which was first identified only in 2005. At the time of my visit, EVACC was down to just one Rabb's frog, so the possibility of saving even a single, Noachian pair had obviously passed. The frog, greenish brown with yellow speckles, was about four inches long, with oversized feet that gave it the look of a gawky teenager. Rabb's fringe-limbed tree frogs lived in the forest above El Valle, and they laid their eggs in tree holes. In an unusual, perhaps even unique arrangement, the male frogs cared for the tadpoles by allowing their young, quite literally, to eat the skin off their backs. Griffith said that he thought there were probably many other amphibian species that had been missed in the initial collecting rush for EVACC and had since vanished; it was hard to say how many, since most of them were probably unknown to science. "Unfortunately," he told me, "we are losing all these amphibians before we even know that they exist."

"Even the regular people in El Valle, they notice it," he said. "They tell me, 'What happened to the frogs? We don't hear them calling anymore.' "

WHEN the first reports that frog populations were crashing began to circulate, a few decades ago, some of the most knowledgeable people in the field were the most skeptical. Amphibians are, after all, among the planet's great survivors. The ancestors of today's frogs crawled out of the water some 400 million years ago, and by 250 million years ago the earliest representatives of what would become the modern amphibian orders—one includes frogs and toads, the second newts and salamanders, and the third weird limbless creatures called caecilians—had evolved. This means that amphibians have been around not just longer than mammals, say, or birds; they have been around since before there were dinosaurs.

Most amphibians—the word comes from the Greek meaning "double life"—are still closely tied to the aquatic realm from which they emerged. (The ancient Egyptians thought that frogs were produced by the coupling of land and water during the annual flooding of the Nile.) Their eggs, which have no shells, must be kept moist in order to develop. There are many frogs that, like the Panamanian golden frog, lay their eggs in streams. There are also frogs that lay them in temporary pools, frogs that lay them underground, and frogs that lay them in nests that they construct out of foam. In addition to frogs that carry their eggs on their backs and in pouches, there are frogs that carry them wrapped like bandages around their legs. Until recently, when both of them went extinct, there were two species of frogs, known as gastric-brooding frogs, that carried their eggs in their stomachs and gave birth to little froglets through their mouths.

Amphibians emerged at a time when all the land on earth was part of a single expanse known as Pangaea. Since the breakup of Pangaea, they've adapted to conditions on every continent except Antarctica. Worldwide, just over seven thousand species have been identified, and while the greatest number are found in the tropical rainforests, there are occasional amphibians, like the sandhill frog of Australia, that can live in the desert, and also amphibians, like the wood frog, that can live above the Arctic Circle. Several common North American frogs, including spring peepers, are able to survive the winter frozen solid, like popsicles. Their extended evolutionary history means that even groups of amphibians that, from a human perspective, seem to be fairly similar may, genetically speaking, be as different from one another as, say, bats are from horses.

David Wake, one of the authors of the article that sent me to Panama, was among those who initially did not believe that amphibians were disappearing. This was back in the mid–nineteen-eighties. Wake's students began returning from frog-collecting trips in the Sierra Nevada empty-handed. Wake remembered from his own student days, in the nineteen-sixties, that frogs in the Sierras had been difficult to avoid. "You'd be walking through meadows, and you'd inadvertently step on them," he told me. "They were just everywhere." Wake assumed that his students were going to the wrong spots, or that they just didn't know how to look. Then a postdoc with several years of collecting experience told him that he couldn't find any amphibians, either. "I said, 'OK, I'll go up with you, and we'll go out to some proven places,'" Wake recalled. "And I took him out to this proven place, and we found like two toads."

Part of what made the situation so mystifying was the geography; frogs seemed to be vanishing not only from populated and disturbed areas but also from relatively pristine places, like the Sierras and the mountains of Central America. In the late nineteen-eighties, an American herpetologist went to the Monteverde Cloud Forest Reserve in northern Costa Rica to study the reproductive habits of golden toads. She spent two field seasons looking; where once the toads had mated in writhing masses, a single male was sighted. (The golden toad, now classified as extinct, was actually a bright tangerine color. It was only very distantly related to the Panamanian golden frog, which, owing to a pair of glands located behind its eyes, is also technically a toad.) Around the same time, in central Costa Rica, biologists noticed that the populations of several endemic frog species had crashed. Rare and highly specialized species were vanishing and so, too, were much more familiar ones. In Ecuador, the Jambato toad, a frequent visitor to backyard gardens, disappeared in a matter of years. And in northeastern

Australia the southern day frog, once one of the most common in the region, could no longer be found.

The first clue to the mysterious killer that was claiming frogs from Queensland to California came—perhaps ironically, perhaps not—from a zoo. The National Zoo, in Washington, D.C., had been successfully raising blue poison-dart frogs, which are native to Suriname, through many generations. Then, more or less from one day to the next, the zoo's tank-bred frogs started dropping. A veterinary pathologist at the zoo took some samples from the dead frogs and ran them through an electron scanning microscope. He found a strange microorganism on the animals' skin, which he eventually identified as a fungus belonging to a group known as chytrids.

Chytrid fungi are nearly ubiquitous; they can be found at the tops of trees and also deep underground. This particular species, though, had never been seen before; indeed, it was so unusual that an entire genus had to be created to accommodate it. It was named *Batrachochytrium dendrobatidis—batrachos* is Greek for "frog"—or Bd for short.

The veterinary pathologist sent samples from infected frogs at the National Zoo to a mycologist at the University of Maine. The mycologist grew cultures of the fungus and then sent some of them back to Washington. When healthy blue poison-dart frogs were exposed to the lab-raised Bd, they sickened. Within three weeks, they were dead. Subsequent research showed that Bd interferes with frogs' ability to take up critical electrolytes through their skin. This causes them to suffer what is, in effect, a heart attack.

EVACC can perhaps best be described as a work-in-progress. The week I spent at the center, a team of American volunteers was also there, helping to construct an exhibit. The exhibit was going to be open to the public, so, for biosecurity purposes, the space had to be isolated and equipped with its own separate entrance. There were holes in the walls where, eventually, glass cases were to be mounted, and around the holes someone had painted a mountain landscape very much like what you would see if you stepped outside and looked up at the hills. The highlight of the exhibit was to be a large case full of Panamanian golden frogs, and the volunteers were trying to construct a three-foot-high concrete waterfall for them. But there were problems with the pumping system and difficulties getting replacement parts in a valley with no hardware store. The volunteers seemed to be spending a lot of time hanging around, waiting.

I spent a lot of time hanging around with them. Like Griffith, all of the volunteers were frog lovers. Several, I learned, were zookeepers who worked with amphibians back in the States. (One told me that frogs had ruined his marriage.) I was moved by the team's dedication, which was the same sort of commitment that had gotten the frogs into the "frog hotel" and then had gotten EVACC up and running, if not entirely completed. But I couldn't help also feeling that there was also something awfully sad about the painted green hills and the fake waterfall.

With almost no frogs left in the forests around El Valle, the case for bringing the animals into EVACC has by now clearly been proved. And yet the longer the frogs spend in the center, the tougher it is to explain what they're doing there. The chytrid fungus, it turns out, does not need amphibians in order to survive. This means that even after it has killed off the animals in an area, it continues to live on, doing whatever it is that chytrid fungi do. Thus, were the golden frogs at EVACC allowed to amble back into the actual hills around El Valle, they would sicken and collapse. (Though the fungus can be destroyed by bleach, it's obviously impossible to disinfect an entire rainforest.) Everyone I spoke to at EVACC told me that the center's goal was to maintain the animals until they could be released to repopulate the forests, and everyone also acknowledged that they couldn't imagine how this would actually be done.

"We've got to hope that somehow it's all going to come

together," Paul Crump, a herpetologist from the Houston Zoo who was directing the stalled waterfall project, told me. "We've got to hope that something will happen, and we'll be able to piece it all together, and it will all be as it once was, which now that I say it out loud sounds kind of stupid."

"The point is to be able to take them back, which every day I see more like a fantasy," Griffith said.

Once chytrid swept through El Valle, it didn't stop; it continued to move east. It has also since arrived in Panama from the opposite direction, out of Colombia. Bd has spread through the highlands of South America and down the eastern coast of Australia, and it has crossed into New Zealand and Tasmania. It has raced through the Caribbean and has been detected in Italy, Spain, Switzerland, and France. In the U.S., it appears to have radiated from several points, not so much in a wavelike pattern as in a series of ripples. At this point, it appears to be, for all intents and purposes, unstoppable.

THE same way acoustical engineers speak of "background noise" biologists talk about "background extinction." In ordinary times—times here understood to mean whole geologic epochs—extinction takes place only very rarely, more rarely even than speciation, and it occurs at what's known as the background extinction rate. This rate varies from one group of organisms to another; often it's expressed in terms of extinctions per million species-years. Calculating the background extinction rate is a laborious task that entails combing through whole databases' worth of fossils. For what's probably the best-studied group, which is mammals, it's been reckoned to be roughly .25 per million species-years. This means that, since there are about fifty-five hundred mammal species wandering around today, at the background extinction rate you'd expect—once again, very roughly—one species to disappear every seven hundred years.

Mass extinctions are different. Instead of a background hum

there's a crash, and disappearance rates spike. Anthony Hallam and Paul Wignall, British paleontologists who have written extensively on the subject, define mass extinctions as events that eliminate a "significant proportion of the world's biota in a geologically insignificant amount of time." Another expert, David Jablonski, characterizes mass extinctions as "substantial biodiversity losses" that occur rapidly and are "global in extent." Michael Benton, a paleontologist who has studied the end-Permian extinction, uses the metaphor of the tree of life: "During a mass extinction, vast swathes of the tree are cut short, as if attacked by crazed, axe-wielding madmen." A fifth paleontologist, David Raup, has tried looking at matters from the perspective of the victims: "Species are at a low risk of extinction most of the time." But this "condition of relative safety is punctuated at rare intervals by a vastly higher risk." The history of life thus consists of "long periods of boredom interrupted occasionally by panic."

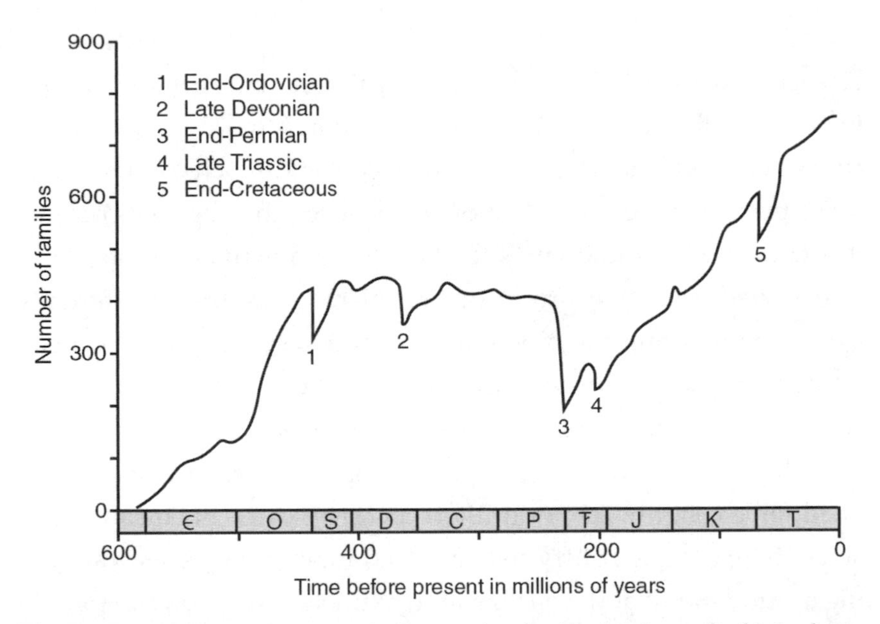

The Big Five extinctions, as seen in the marine fossil record, resulted in a sharp decline in diversity at the family level. If even one species from a family made it through, the family counts as a survivor, so on the species level the losses were far greater.

In times of panic, whole groups of once-dominant organisms can disappear or be relegated to secondary roles, almost as if the globe has undergone a cast change. Such wholesale losses have led paleontologists to surmise that during mass extinction events—in addition to the so-called Big Five, there have been many lesser such events—the usual rules of survival are suspended. Conditions change so drastically or so suddenly (or so drastically *and* so suddenly) that evolutionary history counts for little. Indeed, the very traits that have been most useful for dealing with ordinary threats may turn out, under such extraordinary circumstances, to be fatal.

A rigorous calculation of the background extinction rate for amphibians has not been performed, in part because amphibian fossils are so rare. Almost certainly, though, the rate is lower than it is for mammals. Probably, one amphibian species should go extinct every thousand years or so. That species could be from Africa or from Asia or from Australia. In other words, the odds of an individual's witnessing such an event should be effectively zero. Already, Griffith has observed several amphibian extinctions. Pretty much every herpetologist working out in the field has watched several. (Even I, in the time I spent researching this book, encountered one species that has since gone extinct and three or four others, like the Panamanian golden frog, that are now extinct in the wild.) "I sought a career in herpetology because I enjoy working with animals," Joseph Mendelson, a herpetologist at Zoo Atlanta, has written. "I did not anticipate that it would come to resemble paleontology."

Today, amphibians enjoy the dubious distinction of being the world's most endangered class of animals; it's been calculated that the group's extinction rate could be as much as forty-five thousand times higher than the background rate. But extinction rates among many other groups are approaching amphibian levels. It is estimated that one-third of all reef-building corals, a third of all freshwater mollusks, a third of sharks and rays, a quarter of all mammals, a fifth of all reptiles, and a sixth of all birds are headed toward oblivion. The losses are occurring all over: in the South Pacific and in the North Atlantic, in the Arctic and the Sahel, in lakes and on islands, on mountaintops and in valleys. If you know how to look, you can probably find signs of the current extinction event in your own backyard.

There are all sorts of seemingly disparate reasons that species are disappearing. But trace the process far enough and inevitably you are led to the same culprit: "one weedy species."

Bd is capable of moving on its own. The fungus generates microscopic spores with long, skinny tails; these propel themselves through water and can be carried far longer distances by streams, or in the runoff after a rainstorm. (It's likely this sort of dispersal produced what showed up in Panama as an eastward-moving scourge.) But this kind of movement cannot explain the emergence of the fungus in so many distant parts of the world—Central America, South America, North America, Australia—more or less simultaneously. One theory has it that Bd was moved around the globe with shipments of African clawed frogs, which were used in the nineteen-fifties and sixties in pregnancy tests. (Female African clawed frogs, when injected with the urine of a pregnant woman, lay eggs within a few hours.) Suggestively, African clawed frogs do not seem to be adversely affected by Bd, though they are widely infected with it. A second theory holds that the fungus was spread by North American bullfrogs which have been introduced—sometimes accidentally, sometimes purposefully—into Europe, Asia, and South America, and which are often exported for human consumption. North American bullfrogs, too, are widely infected with Bd but do not seem to be harmed by it. The first has become known as the "Out of Africa" and the second might be called the "frog-leg soup" hypothesis.

Either way, the etiology is the same. Without being loaded by someone onto a boat or a plane, it would have been impossible for a frog carrying Bd to get from Africa to Australia or from North America to Europe. This sort of intercontinental reshuffling, which nowadays we find totally unremarkable, is probably unprecedented in the three-and-a-half-billion-year history of life.

EVEN though Bd has swept through most of Panama by now, Griffith still occasionally goes out collecting for EVACC, looking for survivors. I scheduled my visit to coincide with one of these collecting trips, and one evening I set out with him and two of the American volunteers who were working on the waterfall. We headed east, across the Panama Canal, and spent the night in a region known as Cerro Azul, in a guesthouse ringed by an eight-foot-tall iron fence. At dawn, we drove to the ranger station at the entrance to Chagres National Park. Griffith was hoping to find females of two species that EVACC is short of. He pulled out his government-issued collecting permit and presented it to the sleepy officials manning the station. Some underfed dogs came out to sniff around the truck.

Beyond the ranger station, the road turned into a series of craters connected by deep ruts. Griffith put Jimi Hendrix on the truck's CD player, and we bounced along to the throbbing beat. Frog collecting requires a lot of supplies, so Griffith had hired two men to help with the carrying. At the very last cluster of houses, in the tiny village of Los Ángeles, the men materialized out of the mist. We bounced on until the truck couldn't go any farther; then we all got out and started to walk.

The trail wound its way through the rainforest in a slather of red mud. Every few hundred yards, the main path was crossed by a narrower one; these paths had been made by leaf-cutter ants, making millions—perhaps billions—of trips to bring bits of greenery back to their colonies. (The colonies, which look like mounds of sawdust, can cover an area the size of a city park.) One of the Americans, Chris Bednarski, from the Houston Zoo, warned me to avoid the soldier ants, which will leave their jaws in your shin even after they're dead. "Those'll really mess you up," he observed. The other American, John Chastain, from the Toledo Zoo, was carrying a long hook, for use against venomous snakes. "Fortunately, the ones that can really mess you up are pretty rare," Bednarski assured me. Howler monkeys screamed in the distance. Griffith pointed out jaguar prints in the soft ground.

After about an hour, we came to a farm that someone had carved out of the trees. There was some scraggly corn growing, but no one was around, and it was hard to say whether the farmer had given up on the poor rainforest soil or was simply away for the day. A flock of emerald green parrots shot up into the air. After another several hours, we emerged into a small clearing. A blue morpho butterfly flitted by, its wings the color of the sky. There was a small cabin on the site, but it was so broken down that everyone elected to sleep outside. Griffith helped me string up my bed—a cross between a tent and a hammock that had to be hung between two trees. A slit in the bottom constituted the entryway, and the top was supposed to provide protection against the inevitable rain. When I climbed into the thing, I felt as if I were lying in a coffin.

That evening, Griffith prepared some rice on a portable gas burner. Then we strapped on headlamps and clambered down to a nearby stream. Many amphibians are nocturnal, and the only way to see them is to go looking in the dark, an exercise that's as tricky as it sounds. I kept slipping, and violating Rule No. 1 of rainforest safety: never grab onto something if you don't know what it is. After one of my falls, Bednarski pointed out to me a tarantula the size of my fist sitting on the next tree over.

Practiced hunters can find frogs at night by shining a light into the forest and looking for the reflected glow of their eyes. The first amphibian Griffith sighted this way was a San Jose Cochran frog, perched on top of a leaf. San Jose Cochran frogs are part of a larger family known as "glass frogs," so named because their translucent skin reveals the outline of their internal organs. This particular glass frog was green, with tiny yellow dots. Griffith pulled a pair of surgical gloves out of his pack. He stood completely still and then, with a heronlike gesture, darted to scoop up the frog. With his free hand, he took what looked like the end of a Q-tip and swabbed the frog's belly. He put the Q-tip in a little plastic vial—it would later be sent to a lab and analyzed for Bd—and since it wasn't one of the species he was looking for, he placed the frog back on the leaf. Then he pulled out his camera. The frog stared back at the lens impassively.

We continued to grope through the blackness. Someone spotted a La Loma robber frog, which is orangey-red, like the forest floor; someone else spotted a Warzewitsch frog, which is bright green and shaped like a leaf. With every animal, Griffith went through the same routine: snatching it up, swabbing its belly, photographing it. Finally, we came upon a pair of Panamanian robber frogs locked in amplexus—the amphibian version of sex. Griffith left these two alone.

One of the amphibians that Griffith was hoping to catch, the horned marsupial frog, has a distinctive call that's been likened to the sound of a champagne bottle being uncorked. As we sloshed along—by this point we were walking in the middle of the stream—we heard the call, which seemed to be emanating from several directions at once. At first, it sounded as if it were right nearby, but as we approached, it seemed to get farther away. Griffith began imitating the call, making a cork-popping sound with his lips. Eventually, he decided that the rest of us were scaring the frogs with our splashing. He waded ahead, and we stayed for a long time up to our knees in water, trying not to move. When Griffith finally gestured us over, we found him standing in front of a large yellow frog with long toes and an owlish face. It was sitting on a tree limb, just above eye level. Griffith was looking to find a female horned marsupial frog to add to EVACC's collection. He shot out his arm, grabbed the frog, and flipped it over. Where a female horned marsupial would have a pouch, this one had none. Griffith swabbed it, photographed it, and placed it back in the tree.

"You are a beautiful boy," he murmured to the frog.

Around midnight, we headed back to camp. The only animals that Griffith decided to bring with him were two tiny blue-bellied poison frogs and one whitish salamander, whose species neither he nor the two Americans could identify. The frogs and the salamander were placed in plastic bags with some leaves to keep them moist. It occurred to me that the frogs and their progeny, if they had any, and their progeny's progeny, if they had any, would never again touch the floor of the rainforest but would live out their days in disinfected glass tanks. That night it poured, and in my coffin-like hammock I had vivid, troubled dreams, the only scene from which I could later recall was of a bright yellow frog smoking a cigarette through a holder.

CHAPTER II

THE MASTODON'S MOLARS

Mammut americanum

EXTINCTION MAY BE THE FIRST SCIENTIFIC IDEA THAT KIDS today have to grapple with. One-year-olds are given toy dinosaurs to play with, and two-year-olds understand, in a vague sort of way at least, that these small plastic creatures represent very large animals. If they're quick learners—or, alternatively, slow toilet trainers—children still in diapers can explain that there were once lots of kinds of dinosaurs and that they all died off long ago. (My own sons, as toddlers, used to spend hours over a set of dinosaurs that could be arranged on a plastic mat depicting a forest from the Jurassic or Cretaceous. The scene featured a lava-spewing volcano, which, when you pressed on it, emitted a delightfully terrifying roar.) All of which is to say that extinction strikes us as an obvious idea. It isn't.

Aristotle wrote a ten-book *History of Animals* without ever considering the possibility that animals actually had a history. Pliny's *Natural History* includes descriptions of animals that are real and descriptions of animals that are fabulous, but no descriptions of animals that are extinct. The idea did not crop up during the Middle Ages or during the Renaissance, when the word "fossil" was used to refer to anything dug up from the ground (hence the term "fossil fuels"). In the Enlightenment, the prevailing view was that every species was a link in a great, unbreakable "chain of being." As Alexander Pope put it in his *Essay on Man*:

> All are but parts of one stupendous whole,
> Whose body nature is, and God the soul.

When Carl Linnaeus introduced his system of binomial nomenclature, he made no distinction between the living and the dead because, in his view, none was required. The tenth edition of his *Systema Naturae,* published in 1758, lists sixty-three species of scarab beetle, thirty-four species of cone snail, and sixteen species of flat fishes. And yet in the *Systema Naturae,* there is really only one kind of animal—those that exist.

This view persisted despite a sizable body of evidence to the contrary. Cabinets of curiosities in London, Paris, and Berlin were filled with traces of strange creatures that no one had ever seen—the remains of animals that would now be identified as trilobites, belemnites, and ammonites. Some of the last were so large their fossilized shells approached the size of wagon wheels. In the eighteenth century, mammoth bones increasingly made their way to Europe from Siberia. These, too, were shoehorned into the system. The bones looked a lot like those of elephants. Since there clearly were no elephants in contemporary Russia, it was decided that they must have belonged to beasts that had been washed north in the great flood of Genesis.

Extinction finally emerged as a concept, probably not coincidentally, in revolutionary France. It did so largely thanks to one animal, the creature now called the American mastodon, or *Mammut americanum,* and one man, the naturalist Jean-Léopold-Nicolas-Frédéric Cuvier, known after a dead brother simply as

Georges. Cuvier is an equivocal figure in the history of science. He was far ahead of his contemporaries yet also held many of them back; he could be charming and he could be vicious; he was a visionary and, at the same time, a reactionary. By the middle of the nineteenth century, many of his ideas had been discredited. But the most recent discoveries have tended to support those very theories of his that were most thoroughly vilified, with the result that Cuvier's essentially tragic vision of earth history has come to seem prophetic.

When, exactly, Europeans first stumbled upon the bones of an American mastodon is unclear. An isolated molar unearthed in a field in upstate New York was sent off to London in 1705; it was labeled the "tooth of a Giant." The first mastodon bones subjected to what might, anachronistically, be called scientific study were discovered in 1739. That year, Charles le Moyne, the second Baron de Longueuil, was traveling down the Ohio River with four hundred troops, some, like him, Frenchmen, most of the others Algonquians and Iroquois. The journey was arduous and supplies were short. On one leg, a French soldier would later recall, the troops were reduced to living off acorns. Sometime probably in the fall, Longueuil and his troops set up camp on the east bank of the Ohio, not far from what is now the city of Cincinnati. Several of the Native Americans set off to go hunting. A few miles away, they came to a patch of marsh that gave off a sulfurous smell. Buffalo tracks led to the marsh from all directions, and hundreds—perhaps thousands—of huge bones poked out of the muck, like spars of a ruined ship. The men returned to camp carrying a thigh bone three and a half feet long, an immense tusk, and several huge teeth. The teeth had roots the length of a human hand, and each one weighed nearly ten pounds.

Longueuil was so intrigued by the bones that he instructed his troops to take them along when they broke camp. Lugging the enormous tusk, femur, and molars, the men pushed on through the wilderness. Eventually, they reached the Mississippi River, where they met up with a second contingent of French troops. Over the next several months, many of Longueuil's men died of disease, and the campaign they had come to wage, against the Chickasaw, ended in humiliation and defeat. Nevertheless, Longueuil kept the strange bones safe. He made his way to New Orleans and from there shipped the tusk, the teeth, and the giant femur to France. They were presented to Louis XV, who installed them in his museum, the Cabinet du Roi. Decades later, maps of the Ohio River valley were still largely blank, except for the *Endroit où on a trouvé des os d'Éléphant*—the "place where the elephant bones were found." (Today the "place where the elephant bones were found" is a state park in Kentucky known as Big Bone Lick.)

Longueuil's bones confounded everyone who examined them. The femur and the tusk looked as if they could have belonged to an elephant or, much the same thing according to the taxonomy of the time, a mammoth. But the animal's teeth were a conundrum. They resisted categorization. Elephants' teeth (and also mammoths') are flat on top, with thin ridges that run from side to side, so that the chewing surface resembles the sole of a running shoe. Mastodon teeth, by contrast, are cusped. They do, indeed, look as if they might belong to a jumbo-sized human. The first naturalist to study one of them, Jean-Étienne Guettard, declined even to guess at its provenance.

"What animal does it come from?" he asked plaintively in a paper delivered to France's Royal Academy of Sciences in 1752.

In 1762, the keeper of the king's cabinet, Louis-Jean-Marie Daubenton, tried to resolve the puzzle of the curious teeth by declaring that the "unknown animal of the Ohio" was not an animal at all. Rather, it was two animals. The tusks and leg bones belonged to elephants; the molars came from another creature entirely. Probably, he decided, this other creature was a hippopotamus.

Right around this time, a second shipment of mastodon bones was sent to Europe, this time to London. These remains, also from Big Bone Lick, exhibited the same befuddling pattern: the bones and tusks were elephant-like, while the molars were covered in knobby points. William Hunter, attending physician to the queen, found Daubenton's explanation for the discrepancy wanting. He offered a different explanation—the first halfway accurate one.

"The supposed American elephant," he submitted, was a totally new animal with "which anatomists were unacquainted." It was, he decided, carnivorous, hence its scary-looking teeth. He dubbed the beast the American *incognitum*.

France's leading naturalist, Georges-Louis Leclerc, Comte de Buffon, added yet another twist to the debate. He argued that the remains in question represented not one or two, but three separate animals: an elephant, a hippopotamus, and a third, as-yet-unknown species. With great trepidation, Buffon allowed that this last species—"the largest of them all"—seemed to have disappeared. It was, he proposed, the only land animal ever to have done so.

In 1781, Thomas Jefferson was drawn into the controversy. In his *Notes on the State of Virginia*, written just after he left the state's governorship, Jefferson concocted his own version of the *incognitum*. The animal was, he maintained with Buffon, the largest of all beasts—"five or six times the cubic volume of the elephant." (This would disprove the theory, popular in Europe at the time, that the animals of the New World were smaller and more "degenerate" than those of the Old.) The creature, Jefferson agreed with Hunter, was probably carnivorous. But it was still out there somewhere. If it could not be found in Virginia, it was roaming those parts of the continent that "remain in their aboriginal state, unexplored and undisturbed." When, as president, he dispatched Meriwether Lewis and William Clark to the Northwest, Jefferson hoped that they would come upon live *incognita* roaming its forests.

"Such is the economy of nature," he wrote, "that no instance can be produced of her having permitted any one race of her animals to become extinct; of her having formed any link in her great work so weak as to be broken."

CUVIER arrived in Paris in early 1795, half a century after the remains from the Ohio Valley had reached the city. He was twenty-five years old, with wide-set gray eyes, a prominent nose, and a temperament one friend compared to the exterior of the earth—generally cool but capable of violent tremors and eruptions. Cuvier had grown up in a small town on the Swiss border and had few contacts in the capital. Nevertheless, he had managed to secure a prestigious position there, thanks to the passing of the ancien régime on the one hand and his own sublime self-regard on the other. An older colleague would later describe him as popping up in Paris "like a mushroom."

Cuvier's job at Paris's Museum of Natural History—the democratic successor to the king's cabinet—was, officially, to teach. But in his spare time, he delved into the museum's collection. He spent long hours studying the bones that Longueuil had sent to Louis XV, comparing them with other specimens. On April 4, 1796—or, according to the revolutionary calendar in use at the time, 15 Germinal Year IV—he presented the results of his research at a public lecture.

Cuvier began by discussing elephants. Europeans had known for a long time that there were elephants in Africa, which were considered dangerous, and elephants that resided in Asia, which were said to be more docile. Still, elephants were regarded as elephants, much as dogs were dogs, some gentle and others ferocious. On the basis of his examination of the elephant remains at the museum, including one particularly well-preserved skull from Ceylon and another from the Cape of Good Hope, Cuvier had recognized—correctly, of course—that the two belonged to separate species.

"It is clear that the elephant from Ceylon differs more from that of Africa than the horse from the ass or the goat from the sheep," he declared. Among the animals' many distinguishing characteristics were their teeth. The elephant from Ceylon had molars with wavy ridges on the surface "like festooned ribbons," while the elephant from the Cape of Good Hope had teeth with ridges arranged in the shape of diamonds. Looking at live animals would not have revealed this difference, as who would have the temerity to peer down an elephant's throat? "It is to anatomy alone that zoology owes this interesting discovery," Cuvier declared.

Having successfully, as it were, sliced the elephant in two, Cuvier continued with his dissection. The accepted theory about the giant bones from Russia, Cuvier concluded after "scrupulous examination" of the evidence, was wrong. The teeth and jaws from Siberia "do not exactly resemble those of an elephant." They belonged to another species entirely. As for the teeth of the animal from Ohio, well, a single glance was "sufficient to see that they differ still further."

"What has become of these two enormous animals of which one no longer finds any living traces?" he asked. The question, in Cuvier's formulation, answered itself. They were *espèces perdues*, or lost species. Already, Cuvier had doubled the number of extinct vertebrates, from (possibly) one to two. He was just getting going.

A few months earlier, Cuvier had received sketches of a skeleton that had been discovered on the bank of the Río Luján, west of Buenos Aires. The skeleton—twelve feet long and six feet high—had been shipped to Madrid, where it had been painstakingly reassembled. Working from the sketches, Cuvier had identified its owner—once again, correctly—as some sort of outlandishly oversized sloth. He named it *Megatherium*, meaning "giant beast." Though he had never traveled to Argentina, or, for that matter, anywhere farther than Germany, Cuvier was convinced that *Megatherium* was no longer to be found lumbering along the rivers of South America. It, too, had disappeared. The same was true of the so-called Maastricht animal, whose remains—an enormous, pointy jaw studded with sharklike teeth—had been found in a Dutch quarry. (The Maastricht fossil had recently been seized by the French, who occupied the Netherlands in 1795.)

And if there were four extinct species, Cuvier declared, there must be others. The proposal was a daring one to make given the available evidence. On the basis of a few scattered bones, Cuvier had conceived of a whole new way of looking at life. Species died out. This was not an isolated but a widespread phenomenon.

"All these facts, consistent among themselves, and not opposed by any report, seem to me to prove the existence of a world previous to ours," Cuvier said. "But what was this primitive earth? And what revolution was able to wipe it out?"

SINCE Cuvier's day, the Museum of Natural History has grown into a sprawling institution with outposts all over France. Its main buildings, though, still occupy the site of the old royal gardens in the fifth arrondissement. Cuvier didn't just work at the museum; for most of his adulthood, he also lived on the grounds, in a large stucco house that has since been converted into office space. Next to the house, there's now a restaurant and next to that a menagerie, where, on the day that I visited, some wallabies were sunning themselves on the grass. Across the gardens, there's a large hall that houses the museum's paleontology collection.

Pascal Tassy is a director at the museum who specializes in proboscideans, the group that includes elephants and their lost cousins—mammoths, mastodons, and gomphotheres, to name just a few. I went to visit him because he'd promised to take me to see the very bones Cuvier had handled. I found Tassy in his dimly lit office, in the basement under the paleontology hall, sitting amid a mortuary's worth of old skulls. The walls of the office were decorated with covers from old Tintin comic books. Tassy told me he'd decided to become a paleontologist when he was seven, after reading a Tintin adventure about a dig.

We chatted about proboscideans for a while. "They're a fascinating group," he told me. "For instance, the trunk, which is a change of anatomy in the facial area that is truly extraordinary, it evolved separately five times. Two times—yes, that's surprising. But it happened five times independently! We are forced to accept this by looking at the fossils." So far, Tassy said, some 170 proboscidean species have been identified, going back some fifty-five million years, "and this is far from complete, I am sure."

We headed upstairs, into an annex attached to the back of the paleontology hall like a caboose. Tassy unlocked a small room crowded with metal cabinets. Just inside the door, partially wrapped in plastic, stood what resembled a hairy umbrella stand. This, Tassy explained, was the leg of a woolly mammoth, which had been found, frozen and desiccated, on an island off northern Siberia. When I looked at it more closely, I could see that the skin of the leg had been stitched together, like a moccasin. The hair was a very dark brown and seemed, even after more than ten thousand years, to be almost perfectly preserved.

Tassy opened up one of the metal cabinets and placed the contents on a wooden table. These were the teeth that Longueuil had schlepped down the Ohio River. They were huge and knobby and blackened.

"This is the Mona Lisa of paleontology," Tassy said, pointing to the largest of the group. "The beginning of everything. It's incredible because Cuvier himself made the drawing of this tooth. So he looked at it very carefully." Tassy pointed out to me the original catalog numbers, which had been painted on the teeth in the eighteenth century and were now so faded they could barely be made out.

I picked up the largest tooth in both hands. It was indeed a remarkable object. It was around eight inches long and four across—about the size of a brick and nearly as heavy. The cusps—four sets—were pointy, and the enamel was still largely intact. The roots, as thick as ropes, formed a solid mass the color of mahogany.

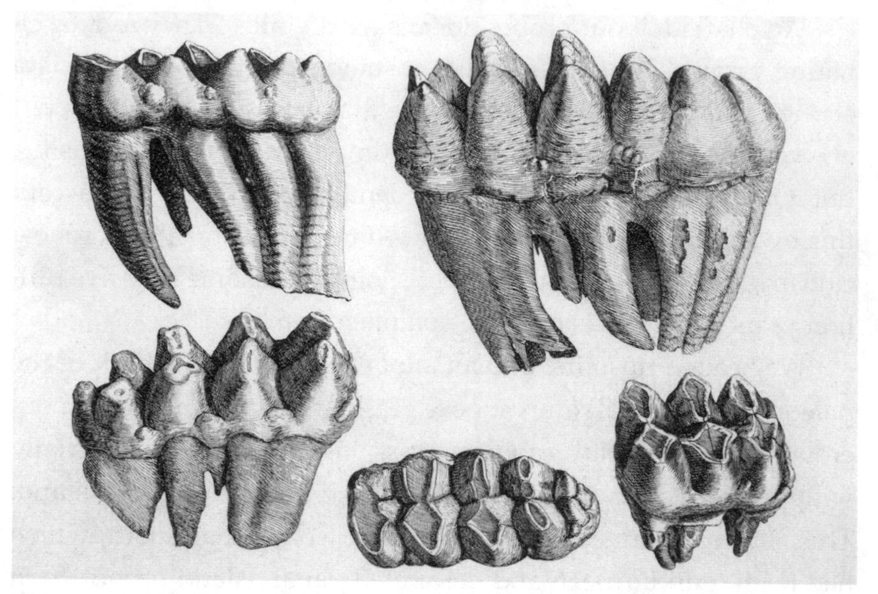

This engraving of mastodon teeth was published with a description by Cuvier in 1812.

From an evolutionary perspective, there's actually nothing strange about a mastodon's molars. Mastodon teeth, like most other mammalian teeth, are composed of a core of dentin surrounded by a layer of harder but more brittle enamel. About thirty million years ago, the proboscidean line that would lead to mastodons split off from the line that would lead to mammoths and elephants. The latter would eventually evolve its more sophisticated teeth, which are made up of enamel-covered plates that have been fused into a shape a bit like a bread loaf. This arrangement is a lot tougher, and it allowed mammoths—and still allows elephants—to consume an unusually abrasive diet. Mastodons, meanwhile, retained their relatively primitive molars (as did humans) and just kept chomping away. Of course, as Tassy pointed out to me, the evolutionary perspective is precisely what Cuvier lacked, which in some ways makes his achievements that much more impressive.

"Sure, he made errors," Tassy said. "But his technical works, most of them are splendid. He was a real fantastic anatomist."

After we had examined the teeth for a while longer, Tassy took me up to the paleontology hall. Just beyond the entrance, the giant femur sent to Paris by Longueuil was displayed, mounted on a pedestal. It was as wide around as a fencepost. French schoolchildren were streaming past us, yelling excitedly. Tassy had a large ring of keys, which he used to open up various drawers underneath the glass display cases. He showed me a mammoth tooth that had been examined by Cuvier and bits of various other extinct species that Cuvier had been the first to identify. Then he took me to look at the Maastricht animal, still today one of the world's most famous fossils. (Though the Netherlands has repeatedly asked for it back, the French have held on to it for more than two hundred years.) In the eighteenth century, the Maastricht fossil was thought by some to belong to a strange crocodile and by others to be from a snaggle-toothed whale. Cuvier would eventually attribute it, yet again correctly, to a marine reptile. (The creature later would be dubbed a mosasaur.)

Around lunchtime, I walked Tassy back to his office. Then I wandered through the gardens to the restaurant next to Cuvier's old house. Because it seemed like the thing to do, I ordered the *Menu Cuvier*—your choice of entrée plus dessert. As I was working my way through the second course—a very tasty cream-filled tart—I began to feel uncomfortably full. I was reminded of a description I had read of the anatomist's anatomy. During the Revolution, Cuvier was thin. In the years he lived on the museum grounds, he grew stouter and stouter, until, toward the end of his life, he became enormously fat.

WITH his lecture on "the species of elephants, both living and fossil," Cuvier had succeeded in establishing extinction as a fact. But his most extravagant assertion—that there had existed a whole lost world, filled with lost species—remained just that. If there had indeed been such a world, traces of other extinct animals ought to be findable. So Cuvier set out to find them.

As it happens, Paris in the seventeen-nineties was a fine place to be a paleontologist. The hills to the north of the city were riddled with quarries that were actively producing gypsum, the main ingredient of plaster of Paris. (The capital grew so haphazardly over so many mines that by Cuvier's day cave-ins were a major hazard.) Not infrequently, miners came upon weird bones, which were prized by collectors, even though they had no real idea what they were collecting. With the help of one such enthusiast, Cuvier had soon assembled the pieces of another extinct animal, which he called *l'animal moyen de Montmartre*—the medium-sized animal from Montmartre.

All the while, Cuvier was soliciting specimens from other naturalists in other parts of Europe. Owing to the reputation the French had earned for seizing objects of value, few collectors would send along actual fossils. But detailed drawings began to arrive from, among other places, Hamburg, Stuttgart, Leiden, and Bologna. "I should say that I have been supported with the most ardent enthusiasm . . . by all Frenchmen and foreigners who cultivate or love the sciences," Cuvier wrote appreciatively.

By 1800, which is to say four years after the elephant paper, Cuvier's fossil zoo had expanded to include twenty-three species he deemed to be extinct. These included: a pygmy hippopotamus, whose remains he discovered in a storeroom at the Paris museum; an elk with enormous antlers whose bones had been found in Ireland; and a large bear—what now would be known as a cave bear—from Germany. The Montmartre animal had, by this point, divided, or multiplied, into six separate species. (Even today, little is known about these species, except that they were ungulates and lived some thirty million years ago.) "If so many lost species have been restored in so little time, how many must be supposed to exist still in the depths of the earth?" Cuvier asked.

Cuvier had a showman's flair and, long before the museum employed public relations professionals, knew how to grab attention. ("He was a man who could have been a star on television today" is how Tassy put it to me.) At one point, the Parisian gypsum mines yielded a fossil of a rabbit-sized creature with a narrow body and a squarish head. Cuvier concluded, based on the shape of its teeth, that the fossil belonged to a marsupial. This was a bold claim, as there were no known marsupials in the Old World. To heighten the drama, Cuvier announced he would put his identification to a public test. Marsupials have a distinctive pair of bones, now known as epipubic bones, that extend from their pelvis. Though these bones were not visible in the fossil as it was presented to him, Cuvier predicted that if he scratched around, the missing bones would be revealed. He invited Paris's scientific elite to gather and watch as he picked away at the fossil with a fine needle. Voilà, the bones appeared. (A cast of the marsupial fossil is on display in Paris in the paleontology hall, but the original is deemed too valuable to be exhibited and so is kept in a special vault.)

Cuvier staged a similar bit of paleontological performance art during a trip to the Netherlands. In a museum in Haarlem, he examined a specimen that consisted of a large, half-moon-shaped skull attached to part of a spinal column. The three-foot-long fossil had been discovered nearly a century earlier and had been attributed—rather curiously, given the shape of the head—to a human. (It had even been assigned a scientific name: *Homo diluvii testis*, or "man who was witness to the Flood.") To rebut this identification, Cuvier first got hold of an ordinary salamander skeleton. Then, with the approval of the Haarlem museum's director, he began chipping away at the rock around the "deluge man's" spine. When he uncovered the fossil animal's forelimbs, they were, just as he had predicted, shaped like a salamander's. The creature was not an antediluvian human but something far weirder: a giant amphibian.

The more extinct species Cuvier turned up, the more the nature of the beasts seemed to change. Cave bears, giant sloths, even giant salamanders—all these bore some relationship to species still alive. But what to make of a bizarre fossil that had been found in a limestone formation in Bavaria? Cuvier received an engraving of this fossil from one of his many correspondents. It showed a tangle of bones, including what looked to be weirdly long arms, skinny fingers, and a narrow beak. The first naturalist to examine it had speculated that its owner had been a sea animal and had used its elongated arms as paddles. Cuvier, on the basis of the engraving, determined—shockingly—that the animal was actually a flying reptile. He called it a *ptero-dactyle,* meaning "wing-fingered."

CUVIER'S discovery of extinction—of "a world previous to ours"—was a sensational event, and news of it soon spread across the Atlantic. When a nearly complete giant skeleton was unearthed by some farmhands in Newburgh, New York, it was recognized as a find of great significance. Thomas Jefferson, at this point the vice president, made several attempts to get his hands on the bones. He failed. But his even more persistent friend, the artist Charles Willson Peale, who'd recently established the nation's first natural history museum, in Philadelphia, succeeded.

Peale, perhaps an even more accomplished showman than Cuvier, spent months fitting together the bones he'd acquired from Newburgh, fashioning the missing pieces out of wood and papier-mâché. He presented the skeleton to the public on Christmas Eve, 1801. To publicize the exhibition, Peale had his black servant, Moses Williams, don an Indian headdress and ride through the streets of Philadelphia on a white horse. The reconstructed beast stood eleven feet high at the shoulder and seventeen feet long from tusks to tail, a somewhat exaggerated size. Visitors were charged fifty cents—quite a considerable sum at the time—for a viewing. The creature—an American mastodon—still lacked an agreed-upon name and was variously referred to as an *incognitum,* an Ohio animal, and, most confusingly of all, a mammoth. It became the world's first blockbuster exhibit and set off a wave of "mammoth

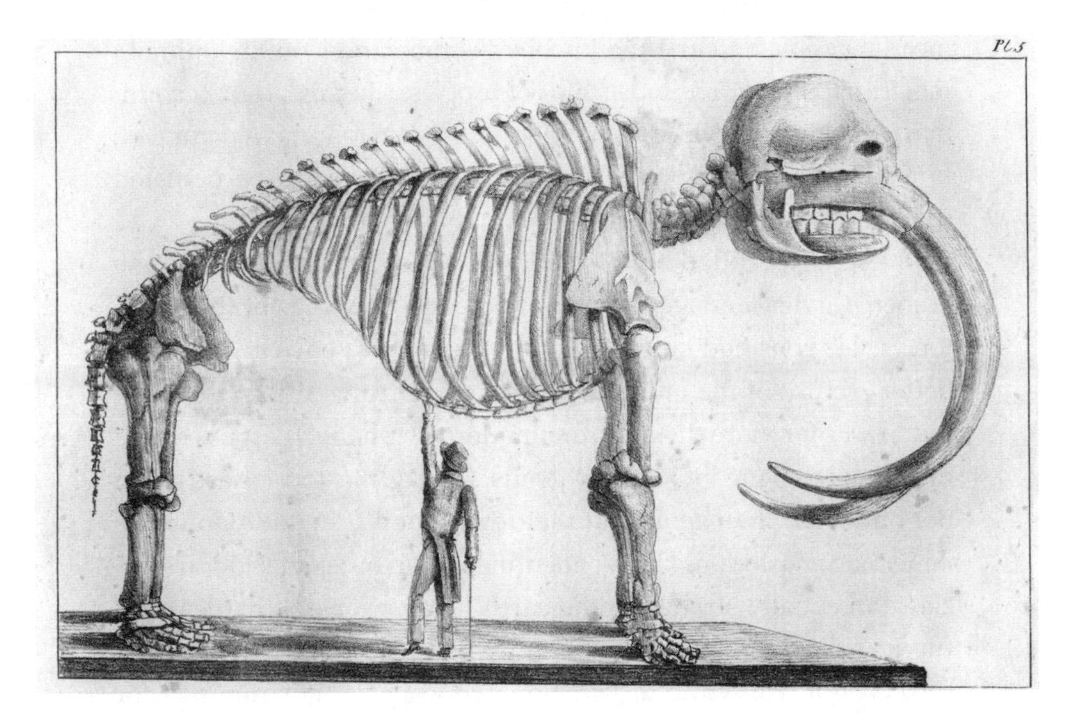

fever." The town of Cheshire, Massachusetts, produced a 1,230-pound "mammoth cheese"; a Philadelphia baker produced a "mammoth bread"; and the newspapers reported on a "mammoth parsnip," a "mammoth peach tree," and a "mammoth eater" who "swallowed 42 EGGS in ten minutes." Peale also managed to piece together a second mastodon, out of additional bones found in Newburgh and nearby towns in the Hudson Valley. After a celebratory dinner held underneath the animal's capacious rib cage, he dispatched this second skeleton to Europe with two of his sons. The skeleton was exhibited for several months in London, during which time the younger Peales decided that the animals' tusks must have pointed downward, like a walrus's. Their plan was to take the skeleton on to Paris and sell it to Cuvier. But while they were still in London, war broke out between Britain and France, making travel between the countries impossible.

Cuvier finally gave the *mastodonte* its name in a paper published in Paris in 1806. The peculiar designation comes from the Greek meaning "breast tooth"; the knobby protuberances on the animal's molars apparently reminded him of nipples. (By this point, the animal had already received a scientific name from a German naturalist; unfortunately this name—*Mammut americanum*—has perpetuated the confusion between mastodons and mammoths.)

Despite the ongoing hostilities between the British and the French, Cuvier managed to obtain detailed drawings of the skeleton Peale's sons had taken to London, and these gave him a much better picture of the animal's anatomy. He realized that the mastodon was far more distant from modern elephants than the mammoth, and assigned it to a new genus. (Today, mastodons are given not only their own genus but their own family.) In addition to the American mastodon, Cuvier identified four other mastodon species, "all equally strange to the earth today." Peale didn't learn of Cuvier's new name until 1809, and when he did, he immediately seized on it. He wrote to Jefferson proposing a "christening" for the mastodon skeleton in his Philadelphia museum. Jefferson was lukewarm about the name Cuvier had come up with—it "may be as good as any other," he sniffed—and didn't deign to respond to the idea of a christening.

In 1812, Cuvier published a four-volume compendium of his work on fossil animals: *Recherches sur les ossemens fossiles de quadrupèdes*. Before he'd begun his "researches," there had been—depending upon who was doing the counting—zero or one extinct vertebrate. Thanks for the most part to his own efforts, there were now forty-nine.

As Cuvier's list grew, so, too, did his renown. Few naturalists dared to announce their findings in public until he had vetted them. "Is not Cuvier the greatest poet of our century?" Honoré de Balzac would ask. "Our immortal naturalist has reconstructed worlds from a whitened bone; rebuilt, like Cadmus, cities from a tooth." Cuvier was honored by Napoleon and, once the Napoleonic Wars finally ended, was invited to Britain, where he was presented at court.

The English were eager converts to Cuvier's project. In the

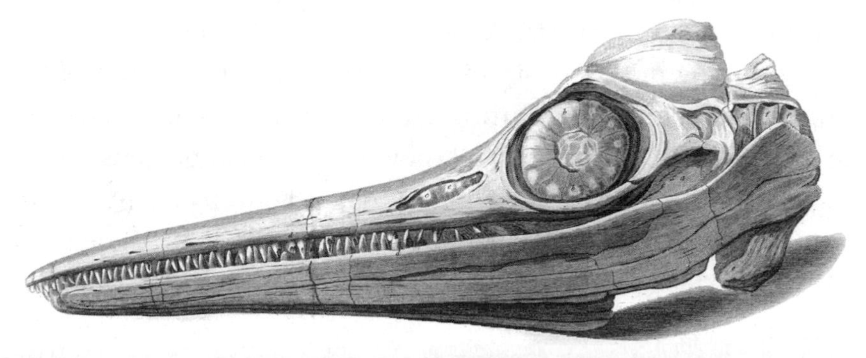

The first ichthyosaur fossil to be discovered was exhibited at London's Egyptian Hall.

early years of the nineteenth century, fossil collecting became so popular among the upper classes that a whole new vocation sprang up. A "fossilist" was someone who made a living hunting up specimens for wealthy patrons. The same year Cuvier published his *Recherches*, one such fossilist, a young woman named Mary Anning, discovered a particularly outlandish specimen. The creature's skull, found in the limestone cliffs of Dorset, was nearly four feet long, with a jaw shaped like a pair of needle-nose pliers. Its eye sockets, peculiarly large, were covered with bony plates.

The fossil ended up in London at the Egyptian Hall, a privately owned museum not unlike Peale's. It was put on exhibit as a fish and then as a relative of a platypus before being recognized as a new kind of reptile—an ichthyosaur, or "fish-lizard." A few years later, other specimens collected by Anning yielded pieces of another, even wilder creature, dubbed a plesiosaur, or "almost-lizard." Oxford's first professor of geology, the Reverend William Buckland, described the plesiosaur as having "the head of a lizard," joined to a neck "resembling the body of a Serpent," the "ribs of a Chameleon, and the paddles of a Whale." Apprised of the find, Cuvier found the account of the plesiosaur so outrageous that he questioned whether the specimens had been doctored. When Anning uncovered another, nearly complete plesiosaur fossil, he was, once again, quickly informed of the finding, at which point he had to acknowledge that he'd been wrong. "One shouldn't anticipate anything more monstrous to emerge," he wrote to one of his English correspondents. During Cuvier's visit to England, he went to visit Oxford, where Buckland showed him yet another astonishing fossil: an enormous jaw with one curved tooth sticking up out of it like a scimitar. Cuvier identified this animal, too, as some sort of lizard. The jaw would, a few decades later, be recognized as belonging to a dinosaur.

The study of stratigraphy was at this point in its infancy, but it was already understood that different layers of rocks had been formed during different periods. The plesiosaur, the ichthyosaur, and the as-yet-unnamed dinosaur had all been found in limestone deposits that were attributed to what was then called the Secondary and is now known as the Mesozoic era. So too, had the *ptero-dactyle* and the Maastricht animal. This pattern led Cuvier to another extraordinary insight about the history of life: it had a direction. Lost species whose remains could be found near the surface of earth, like mastodons and cave bears, belonged to orders of creatures still alive. Dig back farther and one found creatures,

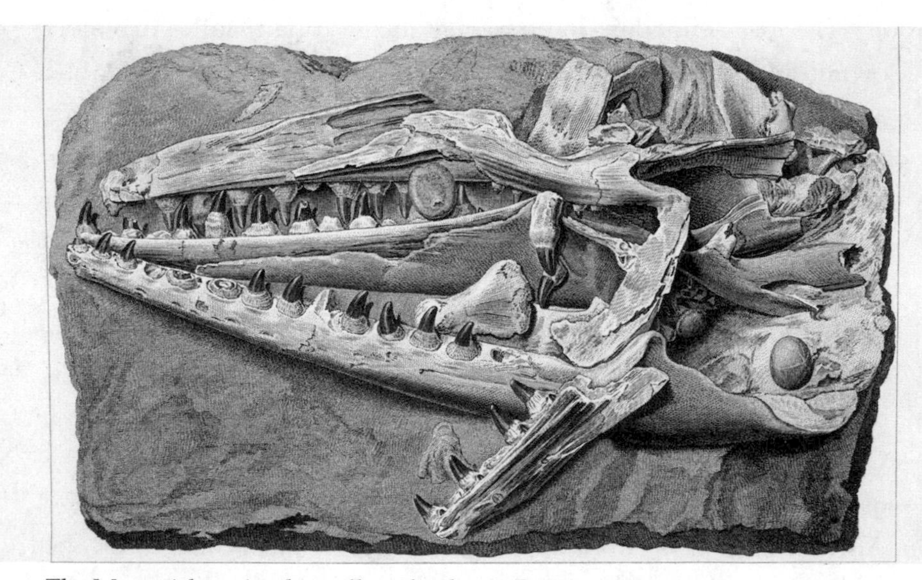

The Maastricht animal is still on display in Paris.

like the animal from Montmartre, that had no obvious modern counterparts. Keep digging and mammals disappeared altogether from the fossil record. Eventually one reached a world not just previous to ours, but a world previous to that, dominated by giant reptiles.

CUVIER'S ideas about this history of life—that it was long, mutable, and full of fantastic creatures that no longer existed—would seem to have made him a natural advocate for evolution. But Cuvier opposed the concept of evolution, or *transformisme* as it was known in Paris at the time, and he tried—generally, it seems, successfully—to humiliate any colleagues who advanced the theory. Curiously, it was the very same skills that led him to discover extinction that made evolution seem to him so preposterous, an affair as unlikely as levitation.

As Cuvier liked to point out, he put his faith in anatomy; this was what had allowed him to distinguish the bones of a mammoth from those of an elephant and to recognize as a giant salamander what others took to be a man. At the heart of his understanding of anatomy was a notion he termed "correlation of parts." By this he meant that the components of an animal all fit together and are optimally designed for its particular way of life; thus, for example, a carnivore will have an intestinal system suited to digesting flesh. At the same time, its jaws will

> be constructed for devouring prey; the claws, for seizing and tearing it, the teeth, for cutting and dividing its flesh; the entire system of its locomotive organs, for pursuing and catching it; its sense organs for detecting it from afar.

Conversely, an animal with hooves must necessarily be an herbivore, since it has "no means of seizing prey." It will have "teeth with a flat crown, to grind seeds and grasses," and a jaw capable of lateral motion. Were any one of these parts to be altered, the functional integrity of the whole would be destroyed. An animal that was born with, say, teeth or sense organs that were somehow different from its parents' would not be able to survive, let alone give rise to a whole new kind of creature.

In Cuvier's day, the most prominent proponent of *transformisme* was his senior colleague at the Museum of Natural History, Jean-Baptiste Lamarck. According to Lamarck, there was a force—the "power of life"—that pushed organisms to become increasingly complex. At the same time, animals and also plants often had to cope with changes in their environment. They did so by adjusting their habits; these new habits, in turn, produced physical modifications that were then passed down to their offspring. Birds that sought prey in lakes spread out their toes when they hit the water, and in this way eventually developed webbed feet and became ducks. Moles, having moved underground, stopped using their sight, and so over generations their eyes became small and weak. Lamarck, for his part, adamantly opposed Cuvier's idea of extinction; there was no process he could imagine capable of wiping an organism out entirely. (Interestingly, the only exception he entertained was humanity, which, Lamarck allowed, might be able to exterminate certain large and slow-to-reproduce animals.) What Cuvier interpreted as *espèces perdues* Lamarck claimed were simply those that had been most completely transformed.

The notion that animals could change their body types when convenient Cuvier found absurd. He lampooned the idea that "ducks by dint of diving became pikes; pikes by dint of happening upon dry land changed into ducks; hens searching for their food at the water's edge, and striving not to get their thighs wet, succeeded so well in elongating their legs that they became herons or storks." He discovered what was, to his mind at least, definitive proof against *transformisme* in a collection of mummies.

When Napoleon had invaded Egypt, the French had, as usual,

seized whatever interested them. Among the crates of loot shipped back to Paris was an embalmed cat. Cuvier examined the mummy, looking for signs of transformation. He found none. The ancient Egyptian cat was, anatomically speaking, indistinguishable from a Parisian alley cat. This proved that species were fixed. Lamarck objected that the few thousand years that had elapsed since the Egyptian cat had been embalmed represented "an infinitely small duration" relative to the vastness of time.

"I know that some naturalists rely a lot on the thousands of centuries that they pile up with a stroke of the pen," Cuvier responded dismissively. Eventually, Cuvier would be called upon to compose a eulogy for Lamarck, which he did very much in the spirit of burying rather than praising. Lamarck, according to Cuvier, was a fantasist. Like the "enchanted palaces of our old romances," his theories were built on "imaginary foundations," so that while they might "amuse the imagination of a poet," they could not "for a moment bear the examination of anyone who has dissected a hand, a viscus, or even a feather."

Having dismissed *transformisme,* Cuvier was left with a gaping hole. He had no account of how new organisms could appear, nor any explanation of how the world could have come to be populated by different groups of animals at different times. This doesn't seem to have bothered him. His interest, after all, was not in the origin of species but in their demise.

THE very first time he spoke about the subject, Cuvier intimated that he knew the driving force behind extinction, if not the exact mechanism. In his lecture on "the species of elephants, both living and fossil," he proposed that the mastodon, the mammoth, and the *Megatherium* had all been wiped out "by some kind of catastrophe." Cuvier hesitated to speculate about the precise nature of this calamity—"It is not for us to involve ourselves in the vast field of conjectures that these questions open up," he said—but he seems to have believed at that point that one disaster would have sufficed.

Later, as his list of extinct species grew, his position changed. There had, he decided, been multiple cataclysms. "Life on earth has often been disturbed by terrible events," he wrote. "Living organisms without number have been the victims of these catastrophes."

Like his view of *transformisme*, Cuvier's belief in cataclysm fit with—indeed, could be said to follow from—his convictions about anatomy. Since animals were functional units, ideally suited to their circumstances, there was no reason why, in the ordinary course of events, they should die out. Not even the most devastating events known to occur in the contemporary world—volcanic eruptions, say, or forest fires—were sufficient to explain extinction; confronted with such changes, organisms simply moved on and survived. The changes that had caused extinctions must therefore have been of a much greater magnitude—so great that animals had been unable to cope with them. That such extreme events had never been observed by him or any other naturalist was another indication of nature's mutability: in the past, it had operated differently—more intensely and more savagely—than it did at present.

"The thread of operations is broken," Cuvier wrote. "Nature has changed course, and none of the agents she employs today would have been sufficient to produce her former works." Cuvier spent several years studying the rock formations around Paris—together with a friend, he produced the first stratigraphic map of the Paris basin—and here, too, he saw signs of cataclysmic change. The rocks showed that the region had, at various points, been submerged. The shifts from one environment to the other—from marine to terrestrial, or, at some points, from marine to freshwater—had, Cuvier decided, "not been slow at all"; rather, they had been brought about by sudden "revolutions on the surface of the earth." The most recent of these revolutions must have occurred relatively recently, for traces of it were still everywhere apparent. This event, Cuvier believed, lay just beyond the edge of recorded history; he observed that many ancient myths and texts, including the Old Testament, allude to some sort of crisis—usually a deluge—that preceded the present order.

Cuvier's ideas about a globe wracked periodically by cataclysm proved very nearly as influential as his original discoveries. His major essay on the subject, which was published in French in 1812, was almost immediately reprinted in English and exported to America. It also appeared in German, Swedish, Italian, Russian, and Czech. But a good deal was lost, or at least misinterpreted, in translation. Cuvier's essay was pointedly secular. He cited the Bible as one of many old (and not entirely reliable) works, alongside the Hindu Vedas and the *Shujing*. This sort of ecumenicalism was unacceptable to the Anglican clergy who made up the faculty at institutions like Oxford, and when the essay was translated into English, it was construed by Buckland and others as offering proof of Noah's flood.

The empirical grounds of Cuvier's theory have, by now, largely been disproved. The physical evidence that convinced him of a "revolution" just prior to recorded history (and that the English interpreted as proof of the Deluge) was, in reality, debris left behind by the last glaciation. The stratigraphy of the Paris basin reflects not sudden "irruptions" of water but rather gradual changes in sea level and the effects of plate tectonics. On all these matters Cuvier was, we now know, wrong.

At the same time, some of Cuvier's most wild sounding claims have turned out to be surprisingly accurate. Life on earth *has* been disturbed by "terrible events," and "organisms without number" have been their victims. Such events cannot be explained by the forces, or "agents," at work in the present. Nature does, on occasion, "change course," and at such moments, it is as if the "thread of operations" has been broken.

Meanwhile, as far as the American mastodon is concerned,

Cuvier was to an almost uncanny extent correct. He decided that the beast had been wiped out five or six thousand years ago, in the same "revolution" that had killed off the mammoth and the *Megatherium*. In fact, the American mastodon vanished around thirteen thousand years ago. Its demise was part of a wave of disappearances that has come to be known as the megafauna extinction. This wave coincided with the spread of modern humans and, increasingly, is understood to have been a result of it. In this sense, the crisis Cuvier discerned just beyond the edge of recorded history was us.

CHAPTER III

THE ORIGINAL PENGUIN

Pinguinus impennis

THE WORD "CATASTROPHIST" WAS COINED IN 1832 BY WILLIAM Whewell, one of the first presidents of the Geological Society of London, who also bequeathed to English "anode," "cathode," "ion," and "scientist." Although the term would later pick up pejorative associations, which stuck to it like burrs, this was not Whewell's intention. When he proposed it, Whewell made it clear that he considered himself a "catastrophist," and that most of the other scientists he knew were catastrophists too. Indeed, there was really only one person he was acquainted with whom the label did not fit, and that was an up-and-coming young geologist named Charles Lyell. For Lyell, Whewell came up with yet another neologism. He called him a "uniformitarian."

Lyell had grown up in the south of England, in the sort of world familiar to fans of Jane Austen. He'd then attended Oxford and trained to become a barrister. Failing eyesight made it difficult for him to practice law, so he turned to the natural sciences instead. As a young man, Lyell made several trips to the Continent and became friendly with Cuvier, at whose house he dined often. He found the older man to be personally "very obliging"—Cuvier allowed him to make casts of several famous fossils to take back with him to England—but Cuvier's vision of earth history Lyell regarded as thoroughly unpersuasive.

When Lyell looked (admittedly myopically) at the rock outcroppings of the British countryside or at the strata of the Paris basin or at the volcanic islands near Naples, he saw no evidence of cataclysm. In fact, quite the reverse: he thought it unscientific (or, as he put it, "unphilosophical") to imagine that change in the world had ever occurred for different reasons or at different rates than it did in the present day. According to Lyell, every feature of the landscape was the result of very gradual processes operating over countless millennia—processes like sedimentation, erosion, and vulcanism, which were all still readily observable. For generations of geology students, Lyell's thesis would be summed up as "The present is the key to the past."

As far as extinction was concerned, this, too, according to Lyell, occurred at a very slow pace—so slow that, at any given time, in any given place, it would not be surprising were it to go unnoticed. The fossil evidence, which seemed to suggest that species had at various points died out en masse, was a sign that the record was unreliable. Even the idea that the history of life had a direction to it—first reptiles, then mammals—was mistaken, another faulty inference drawn from inadequate data. All manner of organisms had existed in all eras, and those that had apparently vanished for good could, under the right circumstances, pop up again. Thus "the huge iguanodon might reappear in the woods, and the ichthyosaur in the sea, while the pterodactyle might flit again through umbrageous groves of tree-ferns." It is clear, Lyell wrote, "that there is no foundation in geological facts for the popular theory of the successive development of the animal and vegetable world."

Lyell published his ideas in three thick volumes, *Principles of Geology: Being an Attempt to Explain the Former Changes of the Earth's*

Awful Changes.
Man found only in a fossil state. —— Reappearance of Ichthyosauri.
"A change came o'er the spirit of my dream." Byron
H.T. de la Beche del & lith. 1830.
A Lecture, — "You will at once perceive," continued Professor Ichthyosaurus, "that the skull before us belonged to some of the lower order of animals the teeth are very insignificant the power of the jaws trifling, and altogether it seems wonderful how the creature could have procured food."

Surface by Reference to Causes Now in Operation. The work was aimed at a general audience, which embraced it enthusiastically. A first print run of forty-five hundred copies quickly sold out, and a second run of nine thousand was ordered up. (In a letter to his fiancée, Lyell boasted that this represented "at least 10 times" as many books as any other English geologist had ever sold.) Lyell became something of a celebrity—the Steven Pinker of his generation—and when he spoke in Boston more than four thousand people tried to get tickets.

For the sake of clarity (and a good read), Lyell had caricatured his opponents, making them sound a great deal more "unphilosophical" than they actually were. They returned the favor. A British geologist named Henry De la Beche, who had a knack for drawing, poked fun at Lyell's ideas about eternal return. He produced a cartoon showing Lyell in the form of a nearsighted ichthyosaur, pointing to a human skull and lecturing to a group of giant reptiles.

"You will at once perceive," Professor Ichthyosaurus tells his pupils in the caption, "that the skull before us belonged to some of the lower order of animals; the teeth are very insignificant, the power of the jaws trifling, and altogether it seems wonderful how the creature could have procured food." De la Beche called the sketch "Awful Changes."

AMONG the readers who snapped up the *Principles* was Charles Darwin. Twenty-two years old and fresh out of Cambridge, Darwin had been invited to serve as a sort of gentleman's companion to the captain of the HMS *Beagle*, Robert FitzRoy. The ship was headed to South America to survey the coast and resolve various mapping discrepancies that hindered navigation. (The Admiralty was particularly interested in finding the best approach to the Falkland Islands, which the British had recently assumed control of.) The voyage, which would last until Darwin was twenty-seven, would take him from Plymouth to Montevideo, through the Strait of Magellan, up to the Galápagos Islands, across the South Pacific to Tahiti, on to New Zealand, Australia, and Tasmania, across the Indian Ocean to Mauritius, around the Cape of Good Hope, and back again to South America. In the popular imagination, the journey is usually seen as the time when Darwin, encountering a wild assortment of giant tortoises, seafaring lizards, and finches with beaks of every imaginable shape and size, discovered natural selection. In fact, Darwin developed his theory only after his return to England, when other naturalists sorted out the jumble of specimens he had shipped back.

It would be more accurate to describe the voyage of the *Beagle* as the period when Darwin discovered Lyell. Shortly before the ship's departure, FitzRoy presented Darwin with a copy of volume one of the *Principles*. Although he was horribly seasick on the first leg of the journey (as he was on many subsequent legs), Darwin reported that he read Lyell "attentively" as the ship headed south. The *Beagle* made its first stop at St. Jago—now Santiago—in the Cape Verde Islands, and Darwin, eager to put his new knowledge to work, spent several days collecting specimens from its rocky cliffs. One of Lyell's central claims was that some areas of the earth were gradually rising, just as others were gradually subsiding. (Lyell further contended that these phenomena were always in balance, so as to "preserve the uniformity of the general relations of the land and sea.") St. Jago seemed to prove his point. The island was clearly volcanic in origin, but it had several curious features, including a ribbon of white limestone halfway up the dark cliffs. The only way to explain these features, Darwin concluded, was as evidence of uplift. The very first place "which I geologised convinced me of the infinite superiority of Lyell's views," he would later write. So taken was Darwin with volume one of the *Principles* that he had volume two shipped to him for pickup at Montevideo. Volume three, it seems, caught up with him in the Falklands.

While the *Beagle* was sailing along the west coast of South

America, Darwin spent several months exploring Chile. He was resting after a hike one afternoon near the town of Valdivia when the ground beneath him began to wobble, as if made of jelly. "One second of time conveys to the mind a strange idea of insecurity, which hours of reflection could never create," he wrote. Several days after the earthquake, arriving in Concepción, Darwin found the entire city had been reduced to rubble. "It is absolutely true, there is not one house left habitable," he reported. The scene was the "most awful yet interesting spectacle" he'd ever witnessed. A series of surveying measurements that FitzRoy took around Concepción's harbor showed that the quake had elevated the beach by nearly eight feet. Once again, Lyell's *Principles* appeared to be rather spectacularly confirmed. Given enough time, Lyell argued, repeated quakes could raise an entire mountain chain many thousands of feet high.

The more Darwin explored the world, the more Lyellian it seemed to him to be. Outside the port of Valparaiso, he found deposits of marine shells far above sea level. These he took to be the result of many episodes of elevation like the one he'd just witnessed. "I have always thought that the great merit of the *Principles* was that it altered the whole tone of one's mind," he would later write. (While in Chile, Darwin also discovered a new and rather remarkable species of frog, which became known as the Chile Darwin's frog. Males of the species incubated their tadpoles in their vocal sacs. Recent searches have failed to turn up any Chile Darwin's frogs, and the species is now believed to be extinct.)

Toward the end of the *Beagle*'s voyage, Darwin encountered coral reefs. These provided him with his first major breakthrough, a startling idea that would ease his entrée into London's scientific circles. Darwin saw that the key to understanding coral reefs was the interplay between biology and geology. If a reef formed around an island or along a continental margin that was slowly sinking, the corals, by growing slowly upward, could maintain their position relative to the water. Gradually, as the land subsided, the corals would form a barrier reef. If, eventually, the land sank away entirely, the reef would form an atoll.

Darwin's account went beyond and to a certain extent contradicted Lyell's; the older man had hypothesized that reefs grew from the rims of submerged volcanoes. But Darwin's ideas were so fundamentally Lyellian in nature that when, upon his return to England, Darwin presented them to Lyell, the latter was delighted. As the historian of science Martin Rudwick has put it, Lyell "recognized that Darwin had out-Lyelled him."

One biographer summed up Lyell's influence on Darwin as follows: "Without Lyell there would have been no Darwin." Darwin himself, after publishing his account of the voyage of the *Beagle* and also a volume on coral reefs, wrote, "I always feel as if my books came half out of Lyell's brains."

LYELL, who saw change occurring always and everywhere in the world around him, drew the line at life. That a species of plant or animal might, over time, give rise to a new one he found unthinkable, and he devoted much of the second volume of the *Principles* to attacking the idea, at one point citing Cuvier's mummified cat experiment in support of his objections.

Lyell's adamant opposition to transmutation, as it was known in London, is almost as puzzling as Cuvier's. New species, Lyell realized, regularly appeared in the fossil record. But how they originated was an issue he never really addressed, except to say that probably each one had begun with "a single pair, or individual, where an individual was sufficient" and multiplied and spread out from there. This process, which seemed to depend on divine or at least occult intervention, was clearly at odds with the precepts he had laid out for geology. Indeed, as one commentator observed, it seemed to require "exactly the kind of miracle" that Lyell had rejected.

With his theory of natural selection, Darwin once again "out-Lyelled" Lyell. Darwin recognized that just as the features of the inorganic world—deltas, river valleys, mountain chains—were brought into being by gradual change, the organic world similarly was subject to constant flux. Ichthyosaurs and plesiosaurs, birds and fish and—most discomfiting of all—humans had come into being through a process of transformation that took place over countless generations. This process, though imperceptibly slow, was, according to Darwin, still very much going on; in biology, as in geology, the present was the key to the past. In one of the most often-quoted passages of *On the Origin of Species,* Darwin wrote:

> It may be said that natural selection is daily and hourly scrutinising, throughout the world, every variation, even the slightest; rejecting that which is bad, preserving and adding up all that is good; silently and insensibly working, whenever and wherever opportunity offers.

Natural selection eliminated the need for any sort of creative miracles. Given enough time for "every variation, even the slightest" to accumulate, new species would emerge from the old. Lyell this time was not so quick to applaud his protégé's work. He only grudgingly accepted Darwin's theory of "descent with modification," so grudgingly that his stance seems to have eventually ruined their friendship.

Darwin's theory about how species originated doubled as a theory of how they vanished. Extinction and evolution were to each other the warp and weft of life's fabric, or, if you prefer, two sides of the same coin. "The appearance of new forms and the disappearance of old forms" were, Darwin wrote, "bound together." Driving both was the "struggle for existence," which rewarded the fit and eliminated the less so.

> The theory of natural selection is grounded on the belief that each new variety, and ultimately each new species, is produced

> and maintained by having some advantage over those with which it comes into competition; and the consequent extinction of less favoured forms almost inevitably follows.

Darwin used the analogy of domestic cattle. When a more vigorous or productive variety was introduced, it quickly supplanted other breeds. In Yorkshire, for example, he pointed out, "it is historically known that the ancient black cattle were displaced by the long-horns," and that these were subsequently "swept away" by the short-horns, "as if by some murderous pestilence."

Darwin stressed the simplicity of his account. Natural selection was such a powerful force that none other was needed. Along with miraculous origins, world-altering catastrophes could be dispensed with. "The whole subject of the extinction of species has been involved in the most gratuitous mystery," he wrote, implicitly mocking Cuvier.

From Darwin's premises, an important prediction followed. If extinction was driven by natural selection and *only* by natural selection, the two processes had to proceed at roughly the same rate. If anything, extinction had to occur more gradually.

"The complete extinction of the species of a group is generally a slower process than their production," he observed at one point.

No one had ever seen a new species produced, nor, according to Darwin, should they expect to. Speciation was so drawn out as to be, for all intents and purposes, unobservable. "We see nothing of these slow changes in progress," he wrote. It stood to reason that extinction should have been that much more difficult to witness. And yet it wasn't. In fact, during the years Darwin spent holed up at Down House, developing his ideas about evolution, the very last individuals of one of Europe's most celebrated species, the great auk, disappeared. What's more, the event was painstakingly chronicled by British ornithologists. Here Darwin's theory was directly contradicted by the facts, with potentially profound implications.

* * *

THE Icelandic Institute of Natural History occupies a new building on a lonely hillside outside Reykjavik. The building has a tilted roof and tilted glass walls and looks a bit like the prow of a ship. It was designed as a research facility, with no public access, which means that a special appointment is needed to see any of the specimens in the institute's collection. These specimens, as I learned on the day of my own appointment, include: a stuffed tiger, a stuffed kangaroo, and a cabinet full of stuffed birds of paradise.

The reason I'd arranged to visit the institute was to see its great auk. Iceland enjoys the dubious distinction of being the bird's last known home, and the specimen I'd come to look at was killed somewhere in the country—no one is sure of the exact spot—in the summer of 1821. The bird's carcass was purchased by a Danish count, Frederik Christian Raben, who had come to Iceland expressly to acquire an auk for his collection (and had nearly drowned in the attempt). Raben took the specimen home to his castle, and it remained in private hands until 1971, when it came up for auction in London. The Institute of Natural History solicited donations, and within three days Icelanders contributed the equivalent of ten thousand British pounds to buy the auk back. (One woman I spoke to, who was ten years old at the time, recalled emptying her piggy bank for the effort.) Icelandair provided two free seats for the homecoming, one for the institute's director and the other for the boxed bird.

Guðmundur Guðmundsson, who's now the institute's deputy director, had been assigned the task of showing me the auk. Guðmundsson is an expert on foraminifera, tiny marine creatures that form intricately shaped shells, known as "tests." On our way to see the bird, we stopped at his office, which was filled with boxes of little glass tubes, each containing a sampling of tests that rattled like sprinkles when I picked it up. Guðmundsson told me that in his spare time he did translating. A few years ago he had completed the first Icelandic rendering of *On the Origin of Species.* He'd found Darwin's prose quite difficult—"sentences inside sentences inside sentences"—and the book, *Uppruni Tegundanna*, had not sold well, perhaps because so many Icelanders are fluent in English.

We made our way to the storeroom for the institute's collection. The stuffed tiger, wrapped in plastic, looked ready to lunge at the stuffed kangaroo. The great auk—*Pinguinus impennis*—was standing off by itself, in a specially made Plexiglas case. It was perched on a fake rock, next to a fake egg.

As the name suggests, the great auk was a large bird; adults grew to be more than two and a half feet tall. The auk could not fly—it was one of the few flightless birds of the Northern Hemisphere—and its stubby wings were almost comically undersized for its body. The auk in the case had brown feathers on its back; probably these were black when the bird was alive but had since faded. "UV light," Guðmundsson said gloomily. "It destroys the plumage." The auk's chest feathers were white, and there was a white spot just beneath each eye. The bird had been stuffed with its most distinctive feature—its large, intricately grooved beak—tipped slightly into the air. This lent it a look of mournful hauteur.

Guðmundsson explained that the great auk had been on display in Reykjavik until 2008, when the institute was restructured by the Icelandic government. At that point, another agency was supposed to create a new home for the bird, but various mishaps, including Iceland's financial crisis, had prevented this from happening, which is why Count Raben's auk was sitting on its fake rock in the corner of the storeroom. On the rock, there was a painted inscription, which Guðmundsson translated for me: THE BIRD WHO IS HERE FOR SHOW WAS KILLED IN 1821. IT IS ONE OF THE FEW GREAT AUKS THAT STILL EXIST.

IN its heyday, which is to say, before humans figured out how to reach its nesting grounds, the great auk ranged from Norway over to Newfoundland and from Italy to Florida, and its population probably numbered in the millions. When the first settlers arrived in Iceland from Scandinavia, great auks were so common that they were regularly eaten for dinner, and their remains have been found in the tenth-century equivalent of household trash. While I was in Reykjavik, I visited a museum built over the ruins of what's believed to be one of the most ancient structures in Iceland—a longhouse constructed out of strips of turf. According to one of the museum's displays, the great auk was "easy prey" for Iceland's medieval inhabitants. In addition to a pair of auk bones, the display featured a video re-creation of an early encounter between man and bird. In the video, a shadowy figure crept along a rocky shore toward a shadowy auk. When he drew close enough, the figure pulled out a stick and clubbed the animal over the head. The auk responded with a cry somewhere between a honk and a grunt. I found the video grimly fascinating and watched it play through a half a dozen times. Creep, clobber, squawk. Repeat.

As best as can be determined, great auks lived much as penguins do. In fact, great auks were the original "penguins." They were called this—the etymology of "penguin" is obscure and may or may not be traced to the Latin *pinguis*, meaning "fat"—by European sailors who encountered them in the North Atlantic. Later, when subsequent generations of sailors met similar-colored flightless birds in the Southern Hemisphere, they used the same name, which led to much confusion, since auks and penguins belong to entirely different families. (Penguins constitute their own family, while auks are members of the family that includes puffins and guillemots; genetic analysis has shown that razorbills are the great auk's closest living relatives.)

Like penguins, great auks were fantastic swimmers—eyewitness accounts attest to the birds' "astonishing velocity" in the water—and they spent most of their lives at sea. But during breeding season, in May and June, they waddled ashore in huge numbers, and here lay their vulnerability. Native Americans clearly hunted the great auk—one ancient grave in Canada was found to contain more than a hundred great auk beaks—as did paleolithic Europeans: great auk bones have been found at archaeological sites in, among other places, Denmark, Sweden, Spain, Italy, and Gibraltar. By the time the first settlers got to Iceland, many of its breeding sites had already been plundered and its range was probably much reduced. Then came the wholesale slaughter.

Lured by the rich cod fishery, Europeans began making regular voyages to Newfoundland in the early sixteenth century. Along the way, they encountered a slab of pinkish granite about fifty acres in area, which rose just above the waves. In the spring, the slab was covered with birds, standing, in a manner of speaking, shoulder to shoulder. Many of these were gannets and guillemots; the rest were great auks. The slab, about forty miles off Newfoundland's northeast coast, became known as the Isle of Birds or, in some accounts, Penguin Island; today it is known as Funk Island. Toward the end of a long transatlantic journey, when provisions were running low, fresh meat was prized, and the ease with which auks could be picked off the slab was soon noted. In an account from 1534, the French explorer Jacques Cartier wrote that some of the Isle of Birds' inhabitants were "as large as geese."

> They are always in the water, not being able to fly in the air, inasmuch as they have only small wings . . . with which . . . they move as quickly along the water as the other birds fly through the air. And these birds are so fat it is marvellous. In less than half an hour we filled two boats full of them, as if they had been stones, so that besides them which we did not eat fresh, every ship did powder and salt five or six barrels full of them.

A British expedition that landed on the island a few years later found it "full of great foules." The men drove a "great number of the foules" into their ships and pronounced the results to be quite tasty—"very good and nourishing meat." A 1622 account by a captain named Richard Whitbourne describes great auks being driven onto boats "by hundreds at a time as if God had made the innocency of so poor a creature to become such an admirable instrument for the sustenation of Man."

Over the next several decades, other uses for the great auk were found besides "sustenation." (As one chronicler observed, "the great auks of Funk Island were exploited in every way that human ingenuity could devise.") Auks were used as fish bait, as a source of feathers for stuffing mattresses, and as fuel. Stone pens were erected on Funk Island—vestiges of these are still visible today—and the birds were herded into the enclosures until someone could find time to butcher them. Or not. According to an English seaman named Aaron Thomas, who sailed to Newfoundland on the HMS *Boston*:

> If you come for their Feathers you do not give yourself the trouble of killing them, but lay hold of one and pluck the best of the Feathers. You then turn the poor Penguin adrift, with his skin half naked and torn off, to perish at his leisure.

There are no trees on Funk Island, and hence nothing to burn. This led to another practice chronicled by Thomas.

> You take a kettle with you into which you put a Penguin or two, you kindle a fire under it, and this fire is absolutely made of the unfortunate Penguins themselves. Their bodys being oily soon produce a Flame.

It's been estimated that when Europeans first landed at Funk Island, they found as many as a hundred thousand pairs of great auks tending to a hundred thousand eggs. (Probably great auks produced only one egg a year; these were about five inches long and speckled, Jackson Pollock–like, in brown and black.) Certainly the island's breeding colony must have been a large one to persist through more than two centuries of depredation. By the late seventeen hundreds, though, the birds' numbers were in sharp decline. The feather trade had become so lucrative that teams of men were spending the entire summer on Funk, scalding and plucking. In 1785, George Cartwright, an English trader and explorer, observed of these teams: "The destruction which they have made is incredible." If a stop were not soon put to their efforts, he predicted, the great auk would soon "be diminished to almost nothing."

Whether the teams actually managed to kill off every last one of the island's auks or whether the slaughter simply reduced the colony to the point that it became vulnerable to other forces is unclear. (Diminishing population density may have made survival less likely for the remaining individuals, a phenomenon that's known as the Allee effect.) In any event, the date that's usually given for the extirpation of the great auk from North America is

Audubon's great auks.

1800. Some thirty years later, while working on *The Birds of America*, John James Audubon traveled to Newfoundland in search of great auks to paint from life. He couldn't find any, and for his illustration had to make do with a stuffed bird from Iceland that had been acquired by a dealer in London. In his description of the great auk, Audubon wrote that it was "rare and accidental on the banks of Newfoundland" and that it was "said to breed on a rock on that island," a curious contradiction since no breeding bird can be said to be "accidental."

ONCE the Funk Island birds had been salted, plucked, and deep-fried into oblivion, there was only one sizable colony of great auks left in the world, on an island called the Geirfuglasker, or great auk skerry, which lay about thirty miles off southwestern Iceland's Reykjanes Peninsula. Much to the auk's misfortune, a volcanic eruption destroyed the Geirfuglasker in 1830. This left the birds one solitary refuge, a speck of an island known as Eldey. By this point, the great auk was facing a new threat: its own rarity. Skins and eggs were avidly sought by gentlemen, like Count Raben, who wanted to fill out their collections. It was in the service of such enthusiasts that the very last known pair of auks was killed on Eldey in 1844.

Before setting out for Iceland, I'd decided that I wanted to see the site of the auk's last stand. Eldey is only about ten miles off the Reykjanes Peninsula, which is just south of Reykjavik. But getting out to the island proved to be way more difficult to arrange than I'd imagined. Everyone I contacted in Iceland told me that no one ever went there. Eventually, a friend of mine who's from Iceland got in touch with his father, who's a minister in Reykjavik, who contacted a friend of his, who runs a nature center in a tiny town on the peninsula called Sandgerði. The head of the nature center, Reynir Sveinsson, in turn, found a fisherman, Halldór Ármannsson, who said he'd be willing to take me, but only if the weather was fair; if it was rainy or windy, the trip would be too dangerous and nausea-inducing, and he wouldn't want to risk it.

Fortunately, the weather on the day we'd fixed turned out to be splendid. I met Sveinsson at the nature center, which features an exhibit on a French explorer, Jean-Baptiste Charcot, who died when his ship, the infelicitously named *Pourquoi-Pas*, sunk off Sandgerði in 1936. We walked over to the harbor and found Ármannsson loading a chest onto his boat, the *Stella*. He explained that inside the chest was an extra life raft. "Regulations," he shrugged. Ármannsson had also brought along his fishing partner and a cooler filled with soda and cookies. He seemed pleased to be making a trip that didn't involve cod.

We motored out of the harbor and headed south, around the Reykjanes Peninsula. It was clear enough that we could see the snow-covered peak of Snæfellsjökull, more than sixty miles away. (To English speakers, Snæfellsjökull is probably best known as the spot where in Jules Verne's *A Journey to the Center of the Earth* the hero finds a tunnel through the globe.) Eldey, being much shorter than Snæfellsjökull, was not yet visible. Sveinsson explained that Eldey's name means "fire island." He said that although he'd spent his entire life in the area, he'd never before been out to it. He'd brought along a fancy camera and was shooting pictures more or less continuously.

As Sveinnson snapped away, I chatted with Ármannsson inside the *Stella*'s small cabin. I was intrigued to see that he had dramatically different colored eyes, one blue and one hazel. Usually, he told me, he fished for cod using a long line that extended six miles and trailed twelve thousand hooks. The baiting of the hooks was his father's job, and it took nearly two days. A good catch could weigh more than seven metric tons. Often Ármannsson slept on the *Stella*, which was equipped with a microwave and two skinny berths.

After a while, Eldey appeared on the horizon. The island looked like the base of an enormous column, or like a giant pedestal waiting for an even more gigantic statue. When we got within maybe a mile, I could see that the top of the island, which from a distance appeared flat, was actually tilted at about a ten-degree angle. We were approaching from the shorter end, so we could look across the entire surface. It was white and appeared to be rippling. As we got closer, I realized that the ripples were birds—so many that they seemed to blanket the island—and when we got even closer, I could see that the birds were gannets—elegant creatures with long necks, cream-colored heads, and tapered beaks. Sveinsson explained that Eldey was home to one of the world's largest colonies of northern gannets—some thirty thousand pairs. He pointed out a pyramid-like structure atop the island. This was a platform for a webcam that Iceland's environmental agency had set up. It was supposed to stream a live feed of the gannets to bird-watchers, but it had not functioned as planned.

"The birds do not like this camera," Sveinsson said. "So they fly over it and shit on it." The guano from thirty thousand gannet pairs has given the island what looks like a coating of vanilla frosting.

Because of the gannets, and perhaps also because of the island's history, visitors are not allowed to step onto Eldey without special (and hard-to-obtain) permits. When I first learned this, I was disappointed, but when we got right up to the island and I saw the way the sea beat against the cliffs, I felt relieved.

THE last people to see great auks alive were around a dozen Icelanders who made the trip to Eldey by rowboat. They set out one evening in June 1844, rowed through the night, and reached the island the following morning. With some difficulty, three of the men managed to clamber ashore at the only possible landing spot: a shallow shelf of rock that extends from the island to the northeast. (A fourth man who was supposed to go with them refused to on the grounds that it was too dangerous.) By this point the island's total auk population, probably never very numerous, appears to have consisted of a single pair of birds and one egg. On catching sight of the humans, the birds tried to run, but they were too slow. Within minutes, the Icelanders had captured the auks and strangled them. The egg, they saw, had been cracked, presumably in the course of the chase, so they left it behind. Two of the men were able to jump back into the boat; the third had to be hauled through the waves with a rope.

The details of the great auks' last moments, including the names of the men who killed the birds—Sigurður Iselfsson, Ketil Ketilsson, and Jón Brandsson—are known because fourteen years later, in the summer of 1858, two British naturalists traveled to Iceland in search of auks. The older of these, John Wolley, was a doctor and an avid egg collector; the younger, Alfred Newton, was a fellow at Cambridge and soon to be the university's first professor of zoology. The pair spent several weeks on the Reykjanes Peninsula, not far from the site of what is now Iceland's international airport, and during that time, they seem to have talked to just about everyone who had ever seen an auk, or even just heard about one, including several of the men who'd made the 1844 expedition.

The pair of birds that had been killed in that outing, they discovered, had been sold to a dealer for the equivalent of about nine pounds. The birds' innards had been sent to the Royal Museum in Copenhagen; no one could say what had happened to the skins. (Subsequent detective work has traced the skin of the female to an auk now on display at the Natural History Museum of Los Angeles.)

Wolley and Newton hoped to get out to Eldey themselves. Wretched weather prevented them. "Boats and men were engaged, and stores laid in, but not a single opportunity occurred when a landing would have been practicable," Newton would later write. "It was with heavy hearts that we witnessed the season wearing away."

Wolley died shortly after the pair returned to England. For Newton, the experience of the trip would prove to be life-altering. He concluded that the auk was gone—"for all practical purposes therefore we may speak of it as a thing of the past"—and he developed what one biographer referred to as a "peculiar attraction" to "extinct and disappearing faunas." Newton realized that the birds

Great auks laid just one egg a year.

that bred along Britain's long coast were also in danger; he noted that they were being gunned down for sport in great numbers.

"The bird that is shot is a parent," he observed in an address to the British Association for the Advancement of Science. "We take advantage of its most sacred instincts to waylay it, and in depriving the parent of life, we doom the helpless offspring to the most miserable of deaths, that by hunger. If this is not cruelty, what is?" Newton argued for a ban on hunting during breeding season, and his lobbying resulted in one of the first laws aimed at what today would be called wildlife protection: the Act for the Preservation of Sea Birds.

As it happens, Darwin's first paper on natural selection appeared in print just as Newton was returning home from Iceland. The paper, in the *Journal of the Proceedings of the Linnean Society*, had—with Lyell's help—been published in a rush soon after Darwin had learned that a young naturalist named Alfred Russel Wallace was onto a similar idea. (A paper by Wallace appeared in the same issue of the *Journal*.) Newton read Darwin's essay very soon after it came out, staying up late into the night to finish it, and he immediately became a convert. "It came to me like the direct revelation of a higher power," he later recalled, "and I awoke next morning with the consciousness that there was an end of all the mystery in the simple phrase, 'Natural Selection.'" He had, he wrote to a friend, developed a case of "pure and unmitigated Darwinism." A few years later, Newton and Darwin became correspondents—at one point Newton sent Darwin a diseased partridge's foot that he thought might be of interest to him—and eventually the two men paid social calls on each other.

Whether the subject of the great auk ever came up in their conversations is unknown. It is not mentioned in Newton and Darwin's surviving correspondence, nor does Darwin allude to the bird or its recent demise in any of his other writings. But Darwin had to be aware of human-caused extinction. In the Galápagos, he had personally witnessed, if not exactly a case of extinction in action, then something very close to it.

Darwin's visit to the archipelago took place in the fall of 1835, nearly four years into the voyage of the *Beagle*. On Charles Island—now Floreana—he met an Englishman named Nicholas Lawson, who was the Galápagos's acting governor as well as the warden of a small, rather miserable penal colony. Lawson was full of useful information. Among the facts he related to Darwin was that on each of the islands in the Galápagos the tortoises had different-shaped shells. On this basis, Lawson claimed that he could "pronounce from which island any tortoise may have been brought." Lawson also told Darwin that the tortoises' days were numbered. The islands were frequently visited by whaling ships, which carried the huge beasts off as portable provisions. Just a few years earlier, a frigate visiting Charles Island had left with two hundred tortoises stowed in its hold. As a result, Darwin noted in his diary, "the numbers have been much reduced." By the time of the *Beagle*'s visit, tortoises had become so scarce on Charles Island that Darwin, it seems, did not see a single one. Lawson predicted that Charles's tortoise, known today by the scientific name *Chelonoidis elephantopus*, would be entirely gone within twenty years. In fact, it probably disappeared in fewer than ten. (Whether *Chelonoidis elephantopus* was a distinct species or a subspecies is still a matter of debate.)

Darwin's familiarity with human-caused extinction is also clear from *On the Origin of Species*. In one of the many passages in which he heaps scorn on the catastrophists, he observes that animals inevitably become rare before they become extinct: "we know this has been the progress of events with those animals which have been exterminated, either locally or wholly, through man's agency." It's a brief allusion and, in its brevity, suggestive. Darwin assumes that his readers are familiar with such "events" and already habituated to them. He himself seems to find nothing remarkable or troubling about this. But human-caused extinction is of course troubling for many reasons, some of which have to do with Darwin's own theory, and it's puzzling that a writer as shrewd and self-critical as Darwin shouldn't have noticed this.

In the *Origin*, Darwin drew no distinction between man and other organisms. As he and many of his contemporaries recognized, this equivalence was the most radical aspect of his work. Humans, just like any other species, were descended, with modification, from more ancient forebears. Even those qualities that seemed to set people apart—language, wisdom, a sense of right and wrong—had evolved in the same manner as other adaptive traits, such as longer beaks or sharper incisors. At the heart of Darwin's theory, as one of his biographers has put it, is "the denial of humanity's special status."

And what was true of evolution should also hold for extinction, since according to Darwin, the latter was merely a side effect of the former. Species were annihilated, just as they were created, by "slow-acting and still existing causes," which is to say, through competition and natural selection; to invoke any other mechanism was nothing more than mystification. But how, then, to make sense of cases like the great auk or the Charles Island tortoise or, to continue the list, the dodo or the Steller's sea cow? These animals had obviously not been done in by a rival species gradually evolving some competitive advantage. They had all been killed off by the same species, and all quite suddenly—in the case of the great auk and the Charles Island tortoise over the course of Darwin's own lifetime. Either there had to be a separate category for human-caused extinction, in which case people really *did* deserve their "special status" as a creature outside of nature, or space in the natural order had to be made for cataclysm, in which case, Cuvier—distressingly—was right.

CHAPTER IV

THE LUCK OF THE AMMONITES

Discoscaphites jerseyensis

THE HILL TOWN OF GUBBIO, ABOUT A HUNDRED MILES NORTH of Rome, might be described as a municipal fossil. Its streets are so narrow that on many of them not even the tiniest Fiat has room to maneuver, and its gray stone piazzas look much as they did in Dante's era. (In fact, it was a powerful Gubbian, installed as lord mayor of Florence, who engineered Dante's exile, in 1302.) If you visit in winter, as I did, when the tourists are gone, the hotels shuttered, and the town's picture-book palace deserted, it almost seems as if Gubbio has fallen under a spell and is waiting to be awoken.

Just beyond the edge of town a narrow gorge leads off to the northeast. The walls of the gorge, which is known as the Gola del Bottaccione, consist of bands of limestone that run in diagonal stripes. Long before people settled the region—long before people existed—Gubbio lay at the bottom of a clear, blue sea. The remains of tiny marine creatures rained down on the floor of that sea, building up year after year, century after century, millennium after millennium. In the uplift that created the Apennine Mountains, the limestone was elevated and tilted at a forty-five-degree angle. To walk up the gorge today is thus to travel, layer by layer, through time. In the space of a few hundred yards, you can cover almost a hundred million years.

The Gola del Bottaccione is now a tourist destination in its own right, though for a more specialized crowd. It is here that in the late nineteen-seventies, a geologist named Walter Alvarez, who had come to study the origins of the Apennines, ended up, more or less by accident, rewriting the history of life. In the gorge, he discovered the first traces of the giant asteroid that ended the Cretaceous period and caused what may have been the worst day ever on planet earth. By the time the dust—in this case, literal as much as figurative—had settled, some three-quarters of all species had been wiped out.

The evidence of the asteroid's impact lies in a thin layer of clay

The clay layer at Gubbio, with a candy marking the spot.

about halfway up the gorge. Sightseers can park at a turnoff constructed nearby. There's also a little kiosk explaining, in Italian, the site's significance. The clay layer is easy to spot. It's been gouged out by hundreds of fingers, a bit like the toes of the bronze St. Peter in Rome, worn down by the kisses of pilgrims. The day I visited was gray and blustery, and I had the place to myself. I wondered what had prompted all that fingering. Was it simple curiosity? A form of geologic rubbernecking? Or was it something more empathetic: the desire to make contact—however attenuated—with a lost world? I, too, of course, had to stick my finger in. I poked around in the groove and scraped out a pebble-sized piece of clay. It was the color of worn brick and the consistency of dried mud. I put it in an old candy wrapper and stuck it in my pocket—my own little chunk of planetary disaster.

WALTER Alvarez came from a long line of distinguished scientists. His great-grandfather and grandfather were both noted physicians, and his father, Luis, was a physicist at the University of California-Berkeley. But it was his mother who took him for long walks in the Berkeley hills and got him interested in geology. Walter attended graduate school at Princeton, then went to work for the oil industry. (He was living in Libya when Muammar Gaddafi took over the country in 1969.) A few years later he got a research post at the Lamont-Doherty Earth Observatory, across the Hudson from Manhattan. At the time, what's sometimes called the "plate tectonics revolution" was sweeping through the profession, and just about everyone at Lamont got swept up in it.

Alvarez decided to try to figure out how, on the basis of plate tectonics, the Italian peninsula had come into being. Key to the project was a kind of reddish limestone, known as the *scaglia rosso*, which can be found, among other places, in the Gola del Bottaccione. The project moved forward, got stuck, and shifted direction. "In science, sometimes it's better to be lucky than smart," he would later say of these events. Eventually, he found himself working in Gubbio with an Italian geologist named Isabella Premoli Silva, who was an expert on foraminifera.

Foraminifera, or "forams" for short, are the tiny marine creatures that create little calcite shells, or tests, which drift down to the ocean floor once the animal inside has died. The tests have a distinctive shape, which varies from species to species; some look (under magnification) like beehives, others like braids or bubbles or clusters of grapes. Forams tend to be widely distributed and abundantly preserved, and this makes them extremely useful as index fossils: on the basis of which species of forams are found in a given layer of rock, an expert like Silva can tell the rock's age. As they worked their way up the Gola del Bottaccione, Silva pointed out to Alvarez a curious sequence. The limestone from the last stage of the Cretaceous period contained diverse, abundant, and relatively large forams, many as big as grains of sand. Directly above that, there was a layer of clay about half an inch thick with no forams in it.

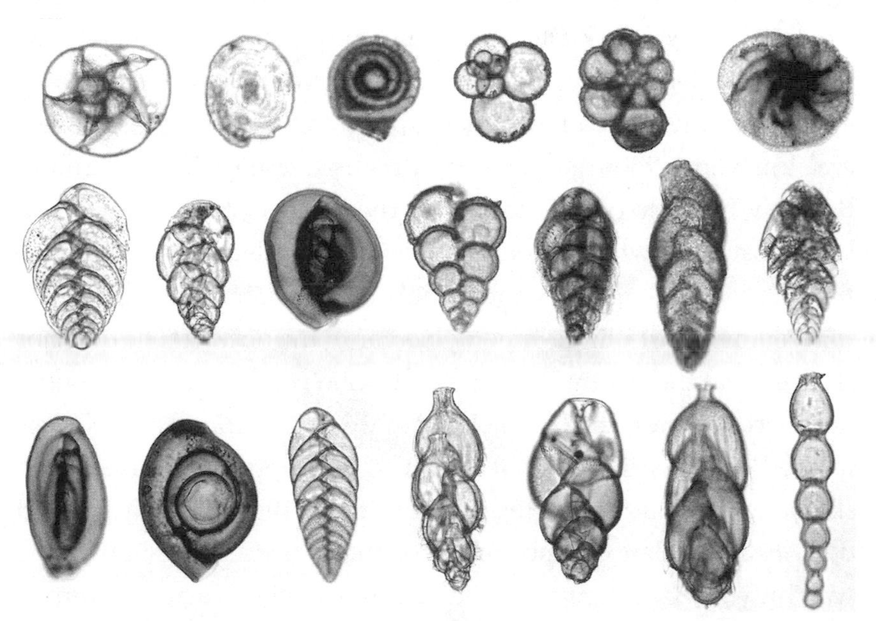

Foraminifera come in distinctive, sometimes whimsical-seeming shapes.

Above the clay there was limestone with more forams, but these belonged to only a handful of species, all of them very tiny and all totally different from the larger ones below.

Alvarez had been schooled in, to use his phrase, a "kind of hard-core uniformitarianism." He'd been trained to believe, after Lyell and Darwin, that the disappearance of any group of organisms had to be a gradual process, with one species slowly dying out, then another, then a third, and so on. Looking at the sequence in the Gubbio limestone, though, he saw something different. The many species of forams in the lower layer seemed to disappear suddenly and all more or less at the same time; the whole process, Alvarez would later recall, certainly "looked very abrupt." Then there was the odd matter of timing. The king-sized forams appeared to vanish right around the point the last of the dinosaurs were known to have died off. This struck Alvarez as more than just a coincidence. He thought it would be interesting to know exactly how much time that half-inch of clay represented.

In 1977, Alvarez got a job at Berkeley, where his father, Luis, was still working, and he brought with him to California his samples from Gubbio. While Walter had been studying plate tectonics, Luis had won a Nobel Prize. He'd also developed the first linear proton accelerator, invented a new kind of bubble chamber, designed several innovative radar systems, and codiscovered tritium. Around Berkeley, Luis had become known as the "wild idea man." Intrigued by a debate over whether there were treasure-filled chambers inside Egypt's second-largest pyramid, he'd at one point designed a test that required installing a muon detector in the desert. (The detector showed that the pyramid was, in fact, solid rock.) At another point, he'd become interested in the Kennedy assassination and had performed an experiment that involved wrapping cantaloupes in shipping tape and shooting them with a rifle. (The experiment demonstrated that the movement of the president's head after he was hit was consistent with the findings of the Warren Commission.) When Walter told his father about the puzzle from Gubbio,

Luis was fascinated. It was Luis who came up with the wild idea of clocking the clay using the element iridium.

Iridium is extremely rare on the surface of the earth but much more common in meteorites. In the form of microscopic grains of cosmic dust, bits of meteorites are constantly raining down on the planet. Luis reasoned that the longer it had taken the clay layer to accumulate, the more cosmic dust would have fallen; thus the more iridium it would contain. He contacted a Berkeley colleague, Frank Asaro, whose lab was one of the few with the right kind of equipment for this sort of analysis. Asaro agreed to run tests on a dozen samples, though he said he very much doubted anything would come of it. Walter gave him some limestone from above the clay layer, some from below it, and some of the clay itself. Then he waited. Nine months later, he got a call. There was something seriously wrong with the samples from the clay layer. The amount of iridium in them was off the charts.

No one knew what to make of this. Was it a weird anomaly, or something more significant? Walter flew to Denmark, to collect some late-Cretaceous sediments from a set of limestone cliffs known as Stevns Klint. At Stevns Klint, the end of the Cretaceous period shows up as a layer of clay that's jet black and smells like dead fish. When the stinky Danish samples were analyzed, they, too, revealed astronomical levels of iridium. A third set of samples, from the South Island of New Zealand, also showed an iridium "spike" right at the end of the Cretaceous.

Luis, according to a colleague, reacted to the news "like a shark smelling blood"; he sensed the opportunity for a great discovery. The Alvarezes batted around theories. But all the ones they could think of either didn't fit the available data or were ruled out by further tests. Then, finally, after almost a year's worth of dead ends, they arrived at the impact hypothesis. On an otherwise ordinary day sixty-five million years ago, an asteroid six miles wide collided with the earth. Exploding on contact, it released energy on the order of a hundred million megatons of TNT, or more than a million of the most powerful H-bombs ever tested. Debris, including iridium from the pulverized asteroid, spread around the globe. Day turned to night, and temperatures plunged. A mass extinction ensued.

The Alvarezes wrote up the results from Gubbio and Stevns Klint and sent them, along with their proposed explanation, to *Science*. "I can remember working very hard to make that paper just as solid as it could possibly be," Walter told me.

THE Alvarezes' paper, "Extraterrestrial Cause for the Cretaceous-Tertiary Extinction," was published in June 1980. It generated lots of excitement, much of it beyond the bounds of paleontology. Journals in disciplines ranging from clinical psychology to herpetology reported on the Alvarezes' findings, and soon the idea of an end-Cretaceous asteroid was picked up by magazines like *Time* and *Newsweek*. One commentator observed that "to connect the dinosaurs, creatures of interest to but the veriest dullards, with a spectacular extraterrestrial event" seemed "like one of those plots a clever publisher might concoct to guarantee sales." Inspired by the impact hypothesis, a group of astrophysicists led by Carl Sagan decided to try to model the effects of an all-out war and came up with the concept of "nuclear winter," which, in turn, generated its own wave of media coverage.

But among professional paleontologists, the Alvarezes' idea and in many cases the Alvarezes themselves were reviled. "The apparent mass extinction is an artifact of statistics and poor understanding of the taxonomy," one paleontologist told the *New York Times*.

"The arrogance of those people is unbelievable," a second asserted. "They know next to nothing about how real animals evolve, live, and become extinct. But despite their ignorance, the geochemists feel that all you have to do is crank up some fancy machine and you've revolutionized science."

"Unseen bolides dropping into an unseen seas are not for me," a third declared.

"The Cretaceous extinctions were gradual and the catastrophe theory is wrong," yet another paleontologist stated. But "simplistic theories will continue to come along to seduce a few scientists and enliven the covers of popular magazines." Curiously enough, the *Times'* editorial board decided to weigh in on the matter. "Astronomers should leave to astrologers the task of seeking the cause of earthly events in the stars," the paper admonished.

To understand the vehemence of this reaction, it helps to go back, once again, to Lyell. In the fossil record, mass extinctions stand out, so much so that the very language that's used to describe earth's history is derived from them. In 1841, John Phillips, a contemporary of Lyell's who succeeded him as president of the Geological Society of London, divided life into three chapters. He called the first the Paleozoic, from the Greek for "ancient life," the second the Mesozoic, meaning "middle life," and the third the Cenozoic, "new life." Phillips fixed as the dividing point between

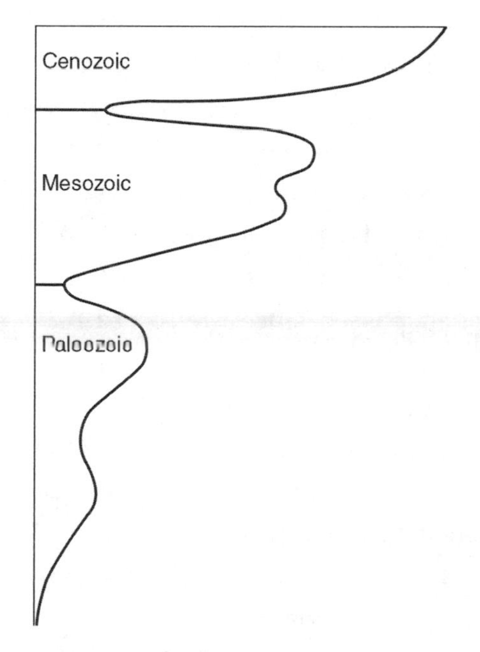

This sketch by John Phillips shows the diversity of life expanding and contracting.

the Paleozoic and the Mesozoic what would now be called the end-Permian extinction, and between the Mesozoic and the Cenozoic, the end-Cretaceous event. (In geologic parlance, the Paleozoic, Mesozoic, and Cenozoic are "eras," and each era comprises several "periods"; the Mesozoic, for example, spans the Triassic, the Jurassic, and the Cretaceous.) The fossils from the three eras were so different that Phillips thought they represented distinct acts of creation.

Lyell was well aware of these breaks in the fossil record. In the third volume of the *Principles of Geology,* he noted a "chasm" between the plants and animals found in rocks from the late Cretaceous period and those found directly above, at the start of the Tertiary period (which is now technically known as the beginning of the Paleogene). For instance, late Cretaceous deposits contained the remains of numerous species of belemnites—squid-like creatures that left behind fossils shaped like bullet casings. But belemnite fossils were never found in more recent deposits. The same pattern held for ammonites, and for rudist bivalves—mollusks that formed immense reefs. (Rudists have been described as oysters pretending to be corals.) To Lyell, it was simply impossible, or "unphilosophical," to imagine that this "chasm" represented what it seemed to—sudden and dramatic global change. So, in a rather neat bit of circular reasoning, he asserted that the faunal gap was just a gap in the fossil record. After comparing the life forms on both sides of the supposed gap, Lyell concluded that the unaccounted-for interval must have been a long one, roughly equivalent to all the time that had passed since the record had resumed. Using today's dating methods, the lacuna he was positing amounts to some sixty-five million years.

Darwin, too, was well informed about the discontinuity at the end of the Cretaceous. In the *Origin,* he observed that the disappearance of the ammonites seemed to be "wonderfully sudden." And, just like Lyell, he dismissed the ammonites and what they seemed to be saying. "For my part," he observed,

> I look at the natural geological record, as a history of the world imperfectly kept, and written in a changing dialect; of this history we possess the last volume alone, relating only to two or three countries. Of this volume, only here and there a short chapter has been preserved; and of each page, only here and there a few lines.

The fragmentary nature of the record meant that the semblance of abrupt change was just that: "With respect to the apparently sudden extermination of whole families or orders," it must be remembered, he wrote, that "wide intervals of time" were probably unaccounted for. Had the evidence of these intervals not been lost, it would have shown "much slow extermination." In this way, Darwin continued the Lyellian project of turning the geologic evidence on its head. "So profound is our ignorance, and so high our presumption, that we marvel when we hear of the extinction of an organic being; and as we do not see the cause, we invoke cataclysms to desolate the world!" he declared.

Darwin's successors inherited the "much slow extermination" problem. The uniformitarian view precluded sudden or sweeping change of any kind. But the more that was learned about the fossil record, the more difficult it was to maintain that an entire age, spanning tens of millions of years, had somehow or other gone missing. This growing tension led to a series of increasingly tortured explanations. Perhaps there *had* been some sort of "crisis" at the close of the Cretaceous, but it had to have been a very slow crisis. Maybe the losses at the end of the period *did* constitute a "mass extinction." But mass extinctions were not to be confused with "catastrophes." The same year that the Alvarezes published their paper in *Science*, George Gaylord Simpson, at the time probably the world's most influential paleontologist, wrote that the "turnover" at the end of the Cretaceous should be regarded as part of "a long and essentially continuous process."

In the context of "hard-core uniformitarianism," the impact

hypothesis was worse than wrong. The Alvarezes were claiming to explain an event that hadn't happened—one that *couldn't* have happened. It was like peddling patent medicine for a fictitious illness. A few years after father and son published their hypothesis, an informal survey was conducted at a meeting of the Society of Vertebrate Paleontology. A majority of those surveyed said they thought some sort of cosmic collision might have taken place. But only one in twenty thought it had anything to do with the extinction of the dinosaurs. One paleontologist at the meeting labeled the Alvarez hypothesis "codswallop."

MEANWHILE, evidence for the hypothesis continued to accumulate.

The first independent corroboration came in the form of tiny grains of rock known as "shocked quartz." Under high magnification, shocked quartz exhibits what look like scratch marks, the result of bursts of high pressure that deform the crystal structure. Shocked quartz was first noted at nuclear test sites and subsequently found in the immediate vicinity of impact craters. In 1984, grains of shocked quartz were discovered in a layer of clay from the Cretaceous-Tertiary, or K-T, boundary in eastern Montana. (*K* is used as the abbreviation for Cretaceous because *C* was already taken by the Carboniferous; today, the border is formally known as the Cretaceous-Paleogene, or K-Pg, boundary.)

The next clue showed up in south Texas, in a curious layer of end-Cretaceous sandstone that seemed to have been produced by an enormous tsunami. It occurred to Walter Alvarez that if there had been a giant, impact-induced tsunami, it would have scoured away shorelines, leaving behind a distinctive fingerprint in the sedimentary record. He scanned the records of thousands of sediment cores that had been drilled in the oceans, and found such a fingerprint in cores from the Gulf of Mexico. Finally, a hundred-mile-wide crater was discovered or, more accurately, rediscovered, beneath the Yucatán Peninsula. Buried under half a mile of newer sediment, the crater had shown up in gravity surveys taken in the nineteen-fifties by Mexico's state-run oil company. Company geologists had interpreted it as the traces of an underwater volcano and, since volcanoes don't yield oil, promptly forgotten about it. When the Alvarezes went looking for cores the company had drilled in the area, they were told that they'd been destroyed in a fire; really, though, they had just been misplaced. The cores were finally located in 1991 and found to contain a layer of glass—rock that had melted, then rapidly cooled—right at the K-T boundary. To the Alvarez camp, this was the clincher, and it was enough to move many uncommitted scientists into the pro-impact column. "Crater supports extinction theory," the *Times* announced. By this point, Luis Alvarez had died of complications from esophageal cancer. Walter dubbed the formation the "Crater of Doom." It became more widely known, after the nearest town, as the Chicxulub crater.

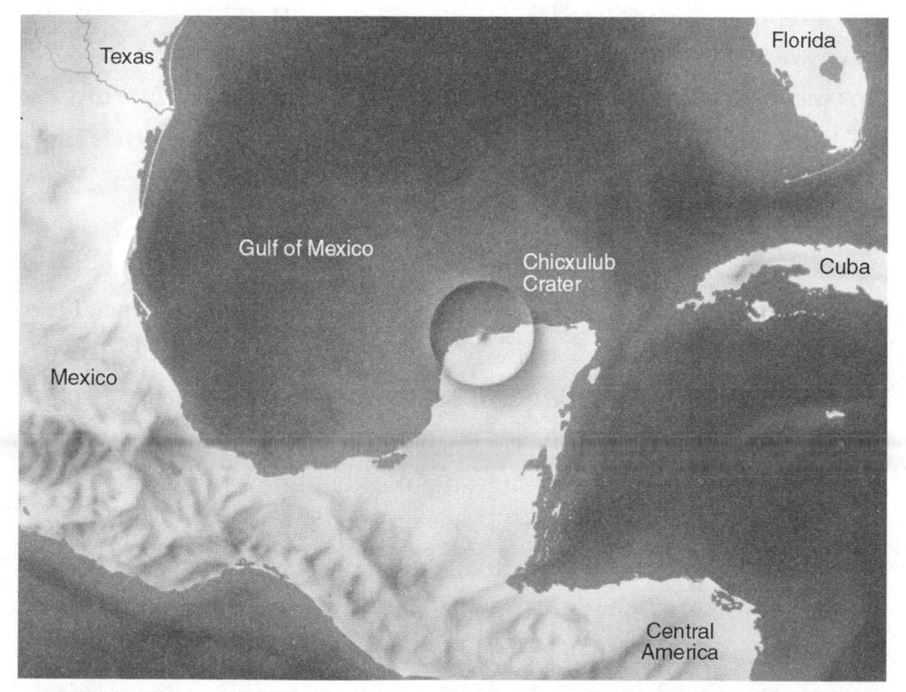

The Chicxulub crater, off the Yucatán Peninsula, is buried under half a mile of sediment.

"Those eleven years seemed long at the time, but looking back they seem very brief," Walter told me. "Just think about it for a moment. Here you have a challenge to a uniformitarian viewpoint that basically every geologist and paleontologist had been trained in, as had their professors and their professors' professors, all the way back to Lyell. And what you saw was people looking at the evidence. And they gradually *did* come to change their minds."

WHEN the Alvarezes published their hypothesis, they knew of only three sites where the iridium layer was exposed: the two Walter had visited in Europe and a third, which they'd been sent samples from, in New Zealand. In the decades since, dozens more have been located, including one near a nude beach in Biarritz, another in the Tunisian desert, and a third in suburban New Jersey. Neil Landman, a paleontologist who specializes in ammonites, often takes field trips to this last site, and one warm fall day I invited myself to tag along. We met in front of the American Museum of Natural History, in Manhattan, where Landman has his office in a turret overlooking Central Park, and, together with a pair of graduate students, headed south to the Lincoln Tunnel.

Driving through northern New Jersey, we passed a succession of strip malls and car dealerships that seemed to repeat every few miles, like dominoes. Eventually, in the general vicinity of Princeton, we pulled into a parking lot next to a baseball field. (Landman would prefer that I not reveal the exact location of the field, for fear of attracting fossil collectors.) In the parking lot, we met up with a geologist named Matt Garb, who teaches at Brooklyn College. Garb, Landman, and the graduate students shouldered their gear. We circumnavigated the baseball field—empty in the middle of a school day—and struck out through the underbrush. Soon we reached a shallow creek. Its banks were covered in rust-colored slime. Brambles hung over the water. Fluttering from these were tattered banners of debris: lost plastic bags, scraps of newspaper, the rings from ancient six-packs. "To me, this is better than Gubbio," Landman announced.

During the late Cretaceous, he explained to me, the park, the creek bed, and everything around us for many miles would have been under water. At that point, the world was very warm—lush forests grew in the Arctic—and sea levels were high. Most of New Jersey formed part of the continental shelf of what's now eastern North America, which, as the Atlantic was then much narrower, was considerably closer to what's now Europe. Landman pointed to a spot in the creek bed a few inches above the water line. There, he told me, was the iridium layer. Although it wasn't in any way visibly different, Landman knew where it was because he'd had the sequence analyzed a few years earlier. Landman is stocky, with a wide face and a graying beard. He had dressed for the trip in khaki shorts and old sneakers. He waded into the creek to join the others, who were already hacking at the bed with their pickaxes. Soon, someone found a fossilized shark's tooth. Someone else dug out a piece of an ammonite. It was about the size of a strawberry and covered in little pimples, or tubercles. Landman identified it as belonging to the species *Discoscaphites iris*.

AMMONITES floated through the world's shallow oceans for more than three hundred million years, and their fossilized shells turn up all around the world. Pliny the Elder, who died in the eruption that buried Pompeii, was already familiar with them, although he considered them to be precious stones. (The stones, he related in his *Natural History*, were said to bring prophetic dreams.) In medieval England, ammonites were known as "serpent stones," and in Germany they were used to treat sick cows. In India, they were—and to a certain extent still are—revered as manifestations of Vishnu.

Like nautiluses, to whom they were distantly related, ammonites constructed spiral shells divided into multiple chambers. The animals themselves occupied only the last and largest chamber; the rest were filled with air, an arrangement that might be compared to an apartment building in which just the penthouse is rented. The walls between the chambers, known as septa, were fantastically elaborate, folded into intricate ruffles, like the edges of a snowflake. (Individual species can be identified by the distinctive patterns of their pleats.) This evolutionary development allowed ammonites to build shells that were at once light and robust—capable of withstanding many atmospheres' worth of water pressure. Most ammonites could fit in a human hand; some grew to be the size of kiddie pools.

Based on the number of teeth ammonites had—nine—it's believed that their closest living kin are octopuses. But since

Ammonite fossils from a nineteenth-century engraving.

ammonites' soft body parts are virtually never preserved, what exactly the animals looked like and how they lived are largely matters of inference. It's probable, though not certain, that they propelled themselves by shooting out a jet of water, which means that they could only travel backward.

"I remember when I was a kid taking paleontology, and I learned that pterodactyls could fly," Landman told me. "My immediate question was, well, how high could they fly? And it's hard to come up with those numbers."

"I've studied ammonites for forty years, and I'm still not sure exactly what they liked," he went on. "I feel they liked water twenty, thirty, maybe forty meters deep. They were swimmers but not very good swimmers. I think they lived a quiet existence." In drawings, ammonites are usually depicted as resembling squids that have been stuffed into snail shells. Landman, however, has trouble with this depiction. He believes that ammonites, though commonly shown with several streaming tentacles, in fact had none. In a drawing that accompanies a recent journal article he published in the journal *Geobios*, ammonites are shown looking like little more than blobs. They have stubby armlike appendages, which are arrayed in a circle and connected by a web of tissue. In males, one of the arms pokes up out of the webbing to form the cephalopod version of a penis.

Landman attended graduate school at Yale in the nineteen-seventies. As a student in the pre-Alvarez days, he was taught that ammonites were declining throughout the Cretaceous, so their eventual disappearance was nothing to get too worked up about. "The sense was, oh, you know, the ammonites were just dying out," he recalled. Subsequent discoveries, many of them made by Landman himself, have shown that, on the contrary, ammonites were doing just fine.

"Here you have lots of species, and we've collected thousands of specimens over the last few years," he told me over the clank of the others' pickaxes. Indeed, in the creek bed, Landman recently came upon two entirely new species of ammonite. One of these he named, in honor of a colleague, *Discoscaphites minardi*. The other he named, in honor of the place, *Discoscaphites jerseyensis*. *Discoscaphites jerseyensis* probably had little spines poking out of its shell, which, Landman speculates, helped the animal appear larger and more intimidating than it actually was.

In their original paper, the Alvarezes proposed that the main cause of the K-T mass extinction was not the impact itself or even the immediate aftermath. The truly catastrophic effect of the asteroid—or, to use the more generic term, bolide—was the dust. In the intervening decades, this account has been subjected to numerous refinements. (The date of the impact has also been pushed back—to sixty-six million years ago.) Though scientists still vigorously argue about many of the details, one version of the event runs as follows:

The bolide arrived from the southeast, traveling at a low angle relative to the earth, so that it came in not so much from above as from the side, like a plane losing altitude. When it slammed into the Yucatán Peninsula, it was moving at something like forty-five thousand miles per hour, and, due to its trajectory, North America was particularly hard-hit. A vast cloud of searing vapor and debris raced over the continent, expanding as it moved and incinerating anything in its path. "Basically, if you were a triceratops in Alberta, you had about two minutes before you got vaporized" is how one geologist put it to me.

In the process of excavating the enormous crater, the asteroid blasted into the air more than fifty times its own mass in pulverized rock. As the ejecta fell back through the atmosphere, the particles incandesced, lighting the sky everywhere at once from directly overhead and generating enough heat to, in effect, broil the surface of the planet. Owing to the composition of the Yucatán Peninsula, the dust thrown up was rich in sulfur. Sulfate aerosols are particularly effective at blocking sunlight, which is the reason a single volcanic eruption, like Krakatoa, can depress global temperatures for years. After the initial heat pulse, the world experienced a multiseason "impact winter." Forests were decimated. Palynologists, who study ancient spores and pollen, have found that diverse plant communities were replaced entirely by rapidly dispersing ferns. (This phenomenon has become known as the "fern spike.") Marine ecosystems effectively collapsed, and they remained in that state for at least half a million, and perhaps as many as several million, years. (The desolate post-impact sea has been dubbed the "Strangelove ocean.")

It's impossible to give anything close to a full account of the various species, genera, families, and even whole orders that went extinct at the K-T boundary. On land, every animal larger than a cat seems to have died out. The event's most famous victims, the dinosaurs—or, to be more precise, the non-avian dinosaurs—suffered a hundred percent losses. Among the groups that were probably alive right up to the end of the Cretaceous were such familiar museum shop fixtures as hadrosaurs, ankylosaurs, tyrannosauruses, and triceratops. (The cover of Walter Alvarez's book on the extinction, *T. Rex and the Crater of Doom*, shows an angry-looking tyrannosaurus reacting with horror to the impact.) Pterosaurs, too, disappeared. Birds were also hard-hit; perhaps three-quarters of all bird families, perhaps more, went extinct. Enantiornithine birds, which retained such archaic features as teeth, were wiped out, as were Hesperornithine birds, which were aquatic and for the most part flightless. The same goes for lizards and snakes; around four-fifths of all species vanished. Mammals' ranks, too, were devastated; something like two-thirds of the mammalian families living at the end of the Cretaceous disappear at the boundary.

In the sea, plesiosaurs, which Cuvier had at first found implausible and then "monstrous," died out. So did mosasaurs, belemnites, and, of course, ammonites. Bivalves, familiar to us today in the form of mussels and oysters, suffered heavy casualties, as did brachiopods, which look like clams but have a totally different anatomy, and bryozoans, which look like corals but once again are totally unrelated. Several groups of marine microorganisms came within a micron or two of annihilation. Among planktonic foraminifera, something like ninety-five percent of all species disappeared, including *Abathomphalus mayaroensis,* whose remains are found in the last layer of Cretaceous limestone in Gubbio. (Planktonic foraminifera live near the ocean surface; benthic species live on the ocean floor.)

In general, the more that's been learned about the K-T boundary, the more wrongheaded Lyell's reading of the fossil record appears. The problem with the record is not that slow extinctions appear abrupt. It's that even abrupt extinctions are likely to look protracted.

Consider the accompanying diagram. Every species has what is known as a "preservation potential"—the odds that an individual of that species will become fossilized—and this varies depending on, among other things, how common the animal is, where it lives, and what it's made out of. (Thick-shelled marine organisms have a much better chance of being preserved than, say, birds with hollow bones.)

In this diagram, the large white circles represent species that are rarely fossilized, the medium-sized circles those that are preserved more frequently, and the small white dots species that are more abundant still. Even if all of these species died out at exactly the same moment, it would *appear* that the white-circle species had vanished much earlier, simply because its remains are rarer. This effect—known as the Signor-Lipps effect, after the scientists who first identified it—tends to "smear out" sudden extinction events, making them look like long, drawn-out affairs.

Following the K-T extinction, it took millions of years for life to recover its former level of diversity. In the meantime, many surviving taxa seem to have shrunk. This phenomenon, which can be seen in the very tiny forams that show up above the iridium layer at Gubbio, is called the Lilliput effect.

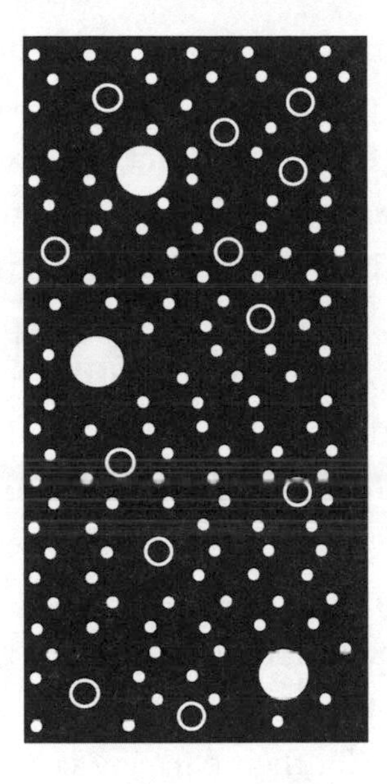

LANDMAN, Garb, and the graduate students chipped away at the creek bed all morning. Although we were in the middle of the country's most densely populated state, not a single person passed by to wonder at what we were doing. As the day grew warmer and more humid, it was pleasant to stand ankle-deep in the water (though I did wonder about the reddish slime). Someone had brought along an empty cardboard box, and, since I didn't have a pickax, I helped out by gathering up the fossils the others had found and arranging them in the box. Several more bits of *Discoscaphites iris* turned up, as well as pieces of an ammonite, *Eubaculites carinatus*, which, instead of having a spiral shell, had one that was long and slender and shaped like a spear. (One theory of the ammonites' demise, popular in the early part of the twentieth century, was that the uncoiled shells of species like *Eubaculites carinatus* indicated that the group had exhausted its practical possibilities and entered some sort of decadent, Lady Gaga-ish phase.) At one point, Garb rushed over in a flurry of excitement. He was carrying a fist-sized chunk of the creek bed and pointed out to me, along one edge, what looked like a tiny fingernail. This, he explained, was a piece of an ammonite's jaw. Ammonite jaws are more common than other body parts but still extremely rare.

"It was worth the trip just for that," he exclaimed.

It's unclear what aspect of the impact—the heat, the darkness, the cold, the change in water chemistry—did in the ammonites. Nor is it entirely clear why some of their cephalopod cousins survived. In contrast to ammonites, nautiluses, for example, sailed through the extinction event: pretty much all of the species known from the end of the Cretaceous survived into the Tertiary.

One theory of the disparity starts with eggs. Ammonites produced very tiny eggs, only a few hundredths of an inch across. The resulting hatchlings, or ammonitellae, had no means of locomotion; they just floated near the surface of the water, drifting along with the current. Nautiluses, for their part, lay very large eggs, among the largest of all invertebrates, nearly an inch in diameter. Hatchling nautiluses emerge, after nearly a year's gestation, as miniature adults and then immediately start swimming around, searching for food in the depths. Perhaps in the aftermath of the impact, conditions at the ocean surface were so toxic that ammonitellae could not survive, while lower down in the water column the situation was less dire, so juvenile nautiluses managed to endure.

Whatever the explanation, the contrasting fate of the two groups raises a key point. Everything (and everyone) alive today is descended from an organism that somehow survived the impact. But it does not follow from this that they (or we) are any better adapted. In times of extreme stress, the whole concept of fitness, at least in a Darwinian sense, loses its meaning: how could a creature be adapted, either well or ill, for conditions it has never before encountered in its entire evolutionary history? At such moments, what Paul Taylor, a paleontologist at London's Natural History Museum, calls "the rules of the survival game" abruptly change. Traits that for many millions of years were advantageous all of a sudden become lethal (though it may be difficult, millions of years after the fact, to identify just what those traits were). And what holds for ammonites and nautiluses applies equally well to belemnites and squids, plesiosaurs and turtles, dinosaurs and mammals. The reason this book is being written by a hairy biped, rather than a scaly one, has more to do with dinosaurian misfortune than with any particular mammalian virtue.

"There's nothing ammonites were doing wrong," Landman told me as we packed up the last fossils from the creek and prepared to head back to New York. "Their hatchlings would have been like plankton, which for all of their existence would have been terrific. What better way to get around and distribute the species? Yet here, in the end, it may well have been their undoing."

CHAPTER V

WELCOME TO THE ANTHROPOCENE

Dicranograptus ziczac

In 1949, a pair of Harvard psychologists recruited two dozen undergraduates for an experiment about perception. The experiment was simple: students were shown playing cards and asked to identify them as they flipped by. Most of the cards were perfectly ordinary, but a few had been doctored, so that the deck contained, among other oddities, a red six of spades and a black four of hearts. When the cards went by rapidly, the students tended to overlook the incongruities; they would, for example, assert that the red six of spades was a six of hearts, or call the black four of hearts a four of spades. When the cards went by more slowly, they struggled to make sense of what they were seeing. Confronted with a red spade, some said it looked "purple" or "brown" or "rusty black." Others were completely flummoxed.

The symbols "look reversed or something," one observed.

"I can't make the suit out, whatever it is," another exclaimed. "I don't know what color it is now or whether it's a spade or heart. I'm not even sure now what a spade looks like! My God!"

The psychologists wrote up their findings in a paper titled "On the Perception of Incongruity: A Paradigm." Among those who found this paper intriguing was Thomas Kuhn. To Kuhn, the twentieth century's most influential historian of science, the experiment was indeed paradigmatic: it revealed how people process disruptive information. Their first impulse is to force it into a familiar framework: hearts, spades, clubs. Signs of mismatch are disregarded for as long as possible—the red spade looks "brown" or "rusty." At the point the anomaly becomes simply too glaring, a crisis ensues—what the psychologists dubbed the "'My God!' reaction."

This pattern was, Kuhn argued in his seminal work, *The Structure of Scientific Revolutions*, so basic that it shaped not only individual perceptions but entire fields of inquiry. Data that did not fit the commonly accepted assumptions of a discipline would either be discounted or explained away for as long as possible. The more contradictions accumulated, the more convoluted the rationalizations became. "In science, as in the playing card experiment, novelty emerges only with difficulty," Kuhn wrote. But then, finally, someone came along who was willing to call a red spade a red spade. Crisis led to insight, and the old framework gave way to a new one. This is how great scientific discoveries or, to use the term Kuhn made so popular, "paradigm shifts" took place.

The history of the science of extinction can be told as a series of paradigm shifts. Until the end of the eighteenth century, the very category of extinction didn't exist. The more strange bones were unearthed—mammoths, *Megatherium*, mosasaurs—the harder naturalists had to squint to fit them into a familiar framework. And squint they did. The giant bones belonged to elephants that had been washed north, or hippos that had wandered west, or whales with malevolent grins. When Cuvier arrived in Paris, he saw that the mastodon's molars could not be fit into the established framework, a "My God" moment that led to him to propose a whole new way of seeing them. Life, Cuvier recognized, had a history. This history was marked by loss and punctuated by events too terrible for human imagining. "Though the world does not change with a change of paradigm, the scientist afterward works in a different world" is how Kuhn put it.

In his *Recherches sur les ossemens fossiles,* Cuvier listed dozens of *espèces perdues,* and he felt sure there were more awaiting discovery. Within a few decades, so many extinct creatures had been identified that Cuvier's framework began to crack. To keep pace with the growing fossil record, the number of disasters had to keep multiplying. "God knows how many catastrophes" would be needed, Lyell scoffed, poking fun at the whole endeavor. Lyell's solution was to reject catastrophe altogether. In Lyell's—and later Darwin's—formulation, extinction was a lonely affair. Each species that had vanished had shuffled off all on its own, a victim of the "struggle for life" and its own defects as a "less improved form."

The uniformitarian account of extinction held up for more than a century. Then, with the discovery of the iridium layer, science faced another crisis. (According to one historian, the Alvarezes' work was "as explosive for science as an impact would have been for earth.") The impact hypothesis dealt with a single moment in time—a terrible, horrible, no-good day at the end of the Cretaceous. But that single moment was enough to crack the framework of Lyell and Darwin. Catastrophes *did* happen.

What is sometimes labeled neocatastrophism, but is mostly nowadays just regarded as standard geology, holds that conditions on earth change only very slowly, except when they don't. In this sense the reigning paradigm is neither Cuvierian nor Darwinian but combines key elements of both—"long periods of boredom interrupted occasionally by panic." Though rare, these moments of panic are disproportionately important. They determine the pattern of extinction, which is to say, the pattern of life.

THE path leads up a hill, across a fast-moving stream, back across the stream, and past the carcass of a sheep, which, more than just dead, looks deflated, like a lost balloon. The hill is bright green but treeless; generations of the sheep's aunts and uncles have kept anything from growing much above muzzle-height. In my view, it's raining. Here in the Southern Uplands of Scotland, though, I'm told by one of the geologists I'm hiking with, this counts only as a light drizzle, or *smirr.*

Our goal is a spot called Dob's Linn, where, according to an old ballad, the Devil himself was pushed over a precipice by a pious shepherd named Dob. By the time we reach the cliff, the *smirr* seems to be smirring harder. There's a view over a waterfall, which crashes down into a narrow valley. A few yards farther up the path there's a jagged outcropping of rock, which is striped vertically, like an umpire's jersey, in bands of light and dark. Jan Zalasiewicz, a stratigrapher from the University of Leicester, sets his rucksack down on the soggy ground and adjusts his red rain jacket. He points to one of the light-colored stripes. "Bad things happened in here," he tells me.

The waterfall at Dob's Linn.

The rocks that we are looking at date back some 445 million years, to the last part of the Ordovician period. At that point, the globe was experiencing a continental logjam; most of the land—including what's now Africa, South America, Australia, and Antarctica—was joined into one giant mass, Gondwana, which spanned more than ninety degrees latitude. England belonged to the continent—now lost—of Avalonia, and Dob's Linn lay in the Southern Hemisphere, at the bottom of an ocean known as the Iapetus.

The Ordovician period followed directly after the Cambrian, which is known, even to the most casual of geology students, for the "explosion" of new life forms that appeared.* The Ordovician, too, was a time when life took off excitedly in new directions—the so-called Ordovician radiation—though it remained, for the most part, still confined to the water. During the Ordovician, the number of marine families tripled, and the seas filled with creatures we would more or less recognize (the progenitors of today's starfish and sea urchins and snails and nautiluses) and also plenty that we would not (conodonts, which probably were shaped like eels; trilobites, which sort of resembled horseshoe crabs; and giant sea scorpions, which, as best as can be determined, looked like something out of a nightmare). The first reefs appeared, and the ancestors of today's clams took on their clam-like form. Toward the middle of the Ordovician, the first plants began to colonize the land. These were very early mosses or liverworts, and they clung low to the ground, as if not quite sure what to make of their new surroundings.

At the end of the Ordovician, some 444 million years ago, the oceans emptied out. Something like eighty-five percent of marine

*A useful mnemonic for remembering the geologic periods of the last half-billion years is: **C**amels **O**ften **S**it **D**own **C**arefully, **P**erhaps **T**heir **J**oints **C**reak (Cambrian-Ordovician-Silurian-Devonian-Carboniferous-Permian-Triassic-Jurassic-Cretaceous). The mnemonic unfortunately runs out before the most recent periods: the Paleogene, the Neogene, and the current Quaternary.

species died off. For a long time, the event was regarded as one of those pseudo-catastrophes that just went to show how little the fossil record could be trusted. Today, it's seen as the first of the Big Five extinctions, and it's thought to have taken place in two brief, intensely deadly pulses. Though its victims are nowhere near as charismatic as those taken out at the end of the Cretaceous, it, too, marks a turning point in life's history—a moment when the rules of the game suddenly flipped, with consequences that, for all intents and purposes, will last forever.

Those animals and plants that made it through the Ordovician extinction "went on to make the modern world," the British paleontologist Richard Fortey has observed. "Had the list of survivors been one jot different, then so would the world today."

ZALASIEWICZ—MY guide at Dob's Linn—is a slight man with shaggy hair, pale blue eyes, and a pleasantly formal manner. He is an expert on graptolites, a once vast and extremely diverse class of marine organisms that thrived during the Ordovician and then, in the extinction event, were very nearly wiped out. To the naked eye, graptolite fossils look like scratches or in some cases tiny petroglyphs. (The word "graptolite" comes from the Greek meaning "written rock"; it was coined by Linnaeus, who dismissed graptolites as mineral encrustations trying to pass themselves off as the remnants of animals.) Viewed through a hand lens, they often prove to have lovely, evocative shapes; one species suggests a feather, another a lyre, a third the frond of a fern. Graptolites were colonial animals; each individual, known as a zooid, built itself a tiny, tubular shelter, known as a theca, which was attached to its neighbor's, like a row house. A single graptolite fossil thus represents a whole community, which drifted or more probably swam along as a single entity, feeding off even smaller plankton. No one knows exactly what the zooids looked like—as with ammonites, the creatures' soft parts resist preservation—but

Graptolite fossils from the early Ordovician.

graptolites are now believed to be related to pterobranchs, a small and hard-to-find class of living marine organisms that resemble Venus flytraps.

Graptolites had a habit—endearing from a stratigrapher's point of view—of speciating, spreading out, and dying off, all in relatively short order. Zalasiewicz compares them to Natasha, the tender heroine of *War and Peace*. They were, he says, "delicate, nervous, and very sensitive to things around them." This makes them useful index fossils—successive species can be used to identify successive layers of rock.

Finding graptolites at Dob's Linn turns out, even for the most amateur of collectors, to be easy. The dark stone in the jagged outcropping is shale. It takes only a gentle hammer-tap to dislodge a chunk. Another tap splits the chunk laterally. It divides like a book opening to a well-thumbed page. Often on the stony surface there's nothing to see, but just as often there's one (or more) faint marks—messages from a former world. One of the graptolites I happen across has been preserved with peculiar clarity. It's shaped like a set of false eyelashes, but very small, as if for a Barbie. Zalasiewicz tells me—doubtless exaggerating—that I have found a "museum quality specimen." I pocket it.

Once Zalasiewicz shows me what to look for, I, too, can make out the arc of the extinction. In the dark shales, graptolites are plentiful and varied. Soon I've collected so many, the pockets of my jacket are sagging. Many of the fossils are variations on the letter *V*, with two arms branching away from a central node. Some look like zippers, others like wishbones. Still others have arms growing off their arms like tiny trees.

The lighter stone, by contrast, is barren. There's barely a graptolite to be found in it. The transition from one state to another—from black stone to gray, from many graptolites to almost none—appears to have occurred suddenly and, according to Zalasiewicz, *did* occur suddenly.

"The change here from black to gray marks a tipping point, if you like, from a habitable sea floor to an uninhabitable one," he tells me. "And one might have seen that in the span of a human lifetime." He describes this transition as distinctly "Cuvierian."

Two of Zalasiewicz's colleagues, Dan Condon and Ian Millar, of the British Geological Survey, have made the hike with us out to Dob's Linn. The pair are experts in isotope chemistry and are planning to collect samples from each of the stripes in the outcropping—samples they hope will contain tiny crystals of zircon. Once back at the lab, they will dissolve the crystals and run the results through a mass spectrometer. This will allow them to say, give or take half a million years or so, when each of the layers was formed. Millar is Scottish and claims to be undaunted by the *smirr*. Eventually, though, even he has to acknowledge that, in English, it's pouring. Rivulets of mud are running down the face of the outcropping, making it impossible to get clean samples. It is decided that we will try again the following day. The three geologists pack up their gear, and we squish back down the trail to the car. Zalasiewicz has made reservations at a bed-and-breakfast in the nearby town of Moffat, whose attractions, I have read, include the world's narrowest hotel and a bronze sheep.

Once everyone has changed into dry clothes, we meet in the sitting room of the B & B for tea. Zalasiewicz has brought along several recent publications of his on graptolites. Settling back in their chairs, Condon and Millar roll their eyes. Zalasiewicz ignores them, patiently explaining to me the import of his latest monograph, "Graptolites in British Stratigraphy," which runs sixty-six single-spaced pages and includes detailed illustrations of more than 650 species. In the monograph, the effects of the extinction show up more systematically, if also less vividly than on the rain-slicked hillside. Until the end of the Ordovician, V-shaped graptolites dominated. These included species like *Dicranograptus ziczac,* whose tiny cups were arranged along arms that curled away and then toward each other, like tusks, and *Adelograptus divergens,* which, in addition to its two main arms, had little side-arms that stuck out like thumbs. Only a handful of graptolite species survived the extinction event; eventually, these diversified and repopulated the seas in the Silurian. But Silurian graptolites had a streamlined body plan, more like a stick than a set of branches. The V-shape had been lost, never to reappear. Here writ very, very small is the fate of the dinosaurs, the mosasaurs, and the ammonites—a once highly successful form relegated to oblivion.

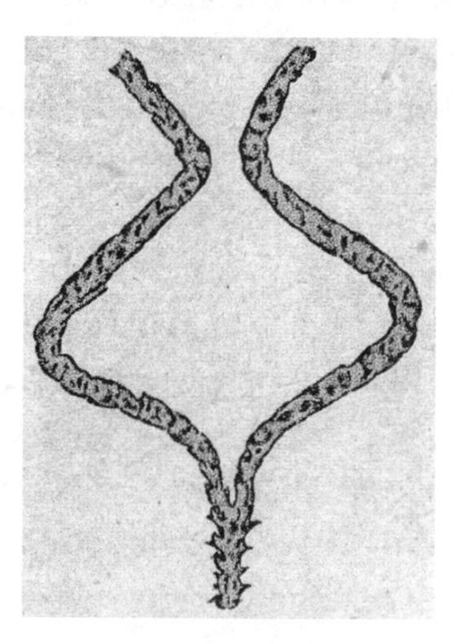

A drawing of the graptolite *Dicranograptus ziczac,* shown several times larger than actual size.

* * *

WHAT happened 444 million years ago to nearly wipe out the graptolites, not to mention the conodonts, the brachiopods, the echinoderms, and the trilobites?

In the years immediately following the publication of the Alvarez hypothesis, it was generally believed—at least among those who considered the hypothesis more than "codswallop"—that a unified theory of mass extinction was at hand. If an asteroid had produced one "chasm" in the fossil record, it seemed reasonable to expect that impacts had caused all of them. This idea received a boost in 1984, when a pair of paleontologists from the University of Chicago published a comprehensive analysis of the marine fossil record. The study revealed that in addition to the five major mass extinctions, there had been many lesser extinction events. When all of these were considered together, a pattern emerged: mass extinctions seemed to take place at regular intervals of roughly twenty-six million years. Extinction, in other words, occurred in periodic bursts, like cicadas crawling out of the earth. The two paleontologists, David Raup and Jack Sepkoski, were unsure what had caused these bursts, but their best guess was some "astronomical and astrophysical cycle," having to do with "the passage of our solar system through the spiral arms of the Milky Way." A group of astrophysicists—as it happened, colleagues of the Alvarezes at Berkeley—took the speculation one step farther. The periodicity, the group argued, could be explained by a small "companion star" to the sun, which, every twenty-six million years, passed through the Oort cloud, producing comet showers that rained destruction on the earth. The fact that no one had ever seen this star, dubbed with horror-movie flair "Nemesis," was, to the Berkeley group, a problem, but not an insurmountable one; there were plenty of small stars out there, still waiting to be cataloged.

In the popular media, what became known as the "Nemesis Affair" generated almost as much excitement as the original asteroid hypothesis. (One reporter described the story as having everything but sex and the royal family.) *Time* ran a cover article, which was soon followed by another disapproving editorial in the *New York Times*. (The editorial pooh-poohed the notion of a "mysterious death-star.") This time, the newspaper was onto something. Though the Berkeley group spent the next year or so scanning the heavens for Nemesis, no glimmer of a "death star" was discovered. More significantly, upon further analysis, the evidence for periodicity began to fall apart. "If there's a consensus, it's that what we were seeing was a statistical fluke," David Raup told me.

Meanwhile, the search for iridium and other signs of extraterrestrial impacts was faltering. Together with many others, Luis Alvarez had thrown himself into this hunt. At a time when scientific collaboration with the Chinese was practically unheard of, he'd managed to obtain rock samples from southern China that spanned the boundary between the Permian and Triassic periods. The end-Permian or Permo-Triassic extinction was the biggest of the Big Five, an episode that came scarily close to eliminating multicellular life altogether. Luis was thrilled to find a layer of clay nestled between the bands of rock from southern China, just as there had been at Gubbio. "We felt sure that there would be lots of iridium there," he would later recall. But the Chinese clay turned out to be, chemically speaking, mundane, its iridium content too infinitesimal to be measured. Higher-than-normal iridium levels were subsequently detected at the end of the Ordovician, in rocks from, among other places, Dob's Linn. However, none of the other telltale signs of an impact, such as shocked quartz, turned up in the right time frame, and it was determined that the elevated iridium levels were more plausibly—if less spectacularly—attributed to the vagaries of sedimentation.

The current theory is that the end-Ordovician extinction was caused by glaciation. For most of the period, a so-called greenhouse climate prevailed—carbon dioxide levels in the air were high and so, too, were sea levels and temperatures. But right around the time of the first pulse of extinction—the one that wreaked havoc among the graptolites—CO_2 levels dropped. Temperatures fell and Gondwana froze. Evidence of the Ordovician glaciation has been found in such far-flung remnants of the supercontinent as Saudi Arabia, Jordan, and Brazil. Sea levels plummeted, and many marine habitats were eliminated, presumably to the detriment of marine organisms. The oceans' chemistry changed, too; among other things, colder water holds more oxygen. No one is sure whether it was the temperature change or one of the many knock-on effects that killed the graptolites; as Zalasiewicz put it to me, "You have a body in the library, and a half a dozen butlers wandering around, looking sheepish." Nor does anyone know what caused the change to begin with. One theory has it that the glaciation was produced by the early mosses that colonized the land and, in so doing, helped draw carbon dioxide out of the air. If this is the case, the first mass extinction of animals was caused by plants.

The end-Permian extinction also seems to have been triggered by a change in the climate. But in this case, the change went in the opposite direction. Right at the time of extinction, 252 million years ago, there was a massive release of carbon into the air—so massive that geologists have a hard time even imagining where all the carbon could have come from. Temperatures soared—the seas warmed by as much as eighteen degrees—and the chemistry of the oceans went haywire, as if in an out-of-control aquarium. The water became acidified, and the amount of dissolved oxygen dropped so low that many organisms probably, in effect, suffocated. Reefs collapsed. The end-Permian extinction took place, though not quite in a human lifetime, in geologic terms nearly as abruptly; according to the latest research by Chinese and American scientists, the whole episode lasted no more than two hundred thousand years, and perhaps less than a hundred thousand. By the time it was over, something like ninety percent of all species on earth had been eliminated. Even intense global warming and ocean acidification seem inadequate to explain losses on such a staggering scale, and so additional mechanisms are still being sought. One hypothesis has it that the heating of the oceans favored bacteria that produce hydrogen sulfide, which is poisonous to most other forms of life. According to this scenario, hydrogen sulfide accumulated in the water, killing off marine creatures, then it leaked into the air, killing off most everything else. The sulfate-reducing bacteria changed the color of the oceans and the hydrogen sulfide the color of the heavens; the science writer Carl Zimmer has described the end-Permian world as a "truly grotesque place" where glassy, purple seas released poisonous bubbles that rose "to a pale green sky."

If twenty-five years ago it seemed that all mass extinctions would ultimately be traced to the same cause, now the reverse seems true. As in Tolstoy, every extinction event appears to be unhappy—and fatally so—in its own way. It may, in fact, be the very freakishness of the events that renders them so deadly; all of a sudden, organisms find themselves facing conditions for which they are, evolutionarily, completely unprepared.

"I think that, after the evidence became pretty strong for the impact at the end of the Cretaceous, those of us who were working on this naively expected that we would go out and find evidence of impacts coinciding with the other events," Walter Alvarez told me. "And it's turned out to be much more complicated. We're seeing right now that a mass extinction can be caused by human beings. So it's clear that we do not have a general theory of mass extinction."

THAT evening in Moffat, once everyone had had enough of tea and graptolites, we went out to the pub on the ground floor of the world's narrowest hotel. After a pint or two, the conversation turned to another one of Zalasiewicz's favorite subjects: giant rats. Rats have followed humans to just about every corner of the globe, and it is Zalasiewicz's professional opinion that one day they will take over the earth.

"Some number will probably stay rat-sized and rat-shaped," he told me. "But others may well shrink or expand. Particularly if there's been epidemic extinction and ecospace opens up, rats may be best placed to take advantage of that. And we know that change in size can take place fairly quickly." I recalled a rat I once watched drag a pizza crust along the tracks at an Upper West Side subway station. I imagined it waddling through a deserted tunnel blown up to the size of a Doberman.

Though the connection might seem tenuous, Zalasiewicz's interest in giant rats represents a logical extension of his interest in graptolites. He is fascinated by the world that preceded humans and also—increasingly—by the world that humans will leave behind. One project informs the other. When he studies the Ordovician, he's trying to reconstruct the distant past on the basis of the fragmentary clues that remain: fossils, isotopes of carbon, layers of sedimentary rock. When he contemplates the future, he's trying to imagine what will remain of the present once the contemporary world has been reduced to fragments: fossils, isotopes of carbon, layers of sedimentary rock. Zalasiewicz is convinced that even a moderately competent stratigrapher will, at the distance of a hundred million years or so, be able to tell that something extraordinary happened at the moment in time that counts for us as today. This is the case even though a hundred million years from now, all that we consider to be the great works of man—the sculptures and the libraries, the monuments and the museums, the cities and the factories—will be compressed into a layer of sediment not much thicker than a cigarette paper. "We have already left a record that is now indelible," Zalasiewicz has written.

One of the ways we've accomplished this is through our restlessness. Often purposefully and just as often not, humans have rearranged the earth's biota, transporting the flora and fauna of Asia to the Americas and of the Americas to Europe and of Europe to Australia. Rats have consistently been on the vanguard of these movements, and they have left their bones scattered everywhere, including on islands so remote that humans never bothered to settle them. The Pacific rat, *Rattus exulans*, a native of southeast Asia, traveled with Polynesian seafarers to, among many other places, Hawaii, Fiji, Tahiti, Tonga, Samoa, Easter Island, and New Zealand. Encountering few predators, stowaway *Rattus exulans* multiplied into what the New Zealand paleontologist Richard Holdaway has described as "a grey tide" that turned "everything edible into rat protein." (A recent study of pollen and animal remains on Easter Island concluded that it wasn't humans who deforested the landscape; rather, it was the rats that came along for the ride and then bred unchecked. The native palms couldn't produce seeds fast enough to keep up with their appetites.) When Europeans arrived in the Americas, and then continued west to the islands the Polynesians had settled, they brought with them the even-more-adaptable Norway rat, *Rattus norvegicus*. In many places, Norway rats, which are actually from China, outcompeted the earlier rat invaders and, in so doing, ravaged the bird and reptile populations the Pacific rats had missed. Rats thus might be said to have created their own "ecospace," which their progeny seem well positioned to dominate. The descendants of today's rats, according to Zalasiewicz, will radiate out to fill the niches that *Rattus exulans* and *Rattus norvegicus* helped empty. He imagines the rats of the future evolving into new shapes and sizes—some "smaller than shrews," others as large as elephants. "We might," he has written, "include among them—for curiosity's sake and to keep our options open—a species or two of large naked rodent, living in caves, shaping rocks as primitive tools and wearing the skins of other mammals that they have killed and eaten."

Meanwhile, whatever the future holds for rats, the extinction event that they are helping to bring about will leave its own distinctive mark. Not yet anywhere near as drastic as the one recorded in the mudstone at Dob's Linn or in the clay layer in Gubbio, it will nevertheless appear in the rocks as a turning point. Climate change—itself a driver of extinction—will also leave behind geologic traces, as will nuclear fallout and river diversion and monoculture farming and ocean acidification.

For all of these reasons, Zalasiewicz believes that we have entered a new epoch, which has no analog in earth's history. "Geologically," he has observed, "this is a remarkable episode."

OVER the years, a number of different names have been suggested for the new age that humans have ushered in. The noted conservation biologist Michael Soulé has suggested that instead of the Cenozoic, we now live in the "Catastrophozoic" era. Michael Samways, an entomologist at South Africa's Stellenbosch University, has floated the term "Homogenocene." Daniel Pauly, a Canadian marine biologist, has proposed the "Myxocene," from the Greek word for "slime," and Andrew Revkin, an American journalist, has offered the "Anthrocene." (Most of these terms owe their origins, indirectly at least, to Lyell, who, back in the eighteen-thirties, coined the words Eocene, Miocene, and Pliocene.)

The word "Anthropocene" is the invention of Paul Crutzen, a Dutch chemist who shared a Nobel Prize for discovering the effects of ozone-depleting compounds. The importance of this discovery is difficult to overstate; had it not been made—and had the chemicals continued to be widely used—the ozone "hole" that opens up every spring over Antarctica would have expanded until eventually it encircled the entire earth. (One of Crutzen's fellow Nobelists reportedly came home from his lab one night and told his wife, "The work is going well, but it looks like it might be the end of the world.")

Crutzen told me that the word "Anthropocene" came to him while he was sitting at a meeting. The meeting's chairman kept referring to the Holocene, the "wholly recent" epoch, which began at the conclusion of the last ice age, 11,700 years ago, and which continues—at least officially—to this day.

"'Let's stop it,'" Crutzen recalled blurting out. "'We are no longer in the Holocene; we are in the Anthropocene.' Well, it was quiet in the room for a while." At the next coffee break, the Anthropocene was the main topic of conversation. Someone came up to Crutzen and suggested that he patent the term.

Crutzen wrote up his idea in a short essay, "Geology of Mankind," that ran in *Nature*. "It seems appropriate to assign the term 'Anthropocene' to the present, in many ways human-dominated, geological epoch," he observed. Among the many geologic-scale changes people have effected, Crutzen cited the following:

- Human activity has transformed between a third and a half of the land surface of the planet.
- Most of the world's major rivers have been dammed or diverted.
- Fertilizer plants produce more nitrogen than is fixed naturally by all terrestrial ecosystems.
- Fisheries remove more than a third of the primary production of the oceans' coastal waters.
- Humans use more than half of the world's readily accessible fresh water runoff.

Most significantly, Crutzen said, people have altered the composition of the atmosphere. Owing to a combination of fossil fuel combustion and deforestation, the concentration of carbon dioxide in the air has risen by forty percent over the last two centuries, while the concentration of methane, an even more potent greenhouse gas, has more than doubled.

"Because of these anthropogenic emissions," Crutzen wrote, the global climate is likely to "depart significantly from natural behavior for many millennia to come."

Crutzen published "Geology of Mankind" in 2002. Soon, the "Anthropocene" began migrating out into other scientific journals.

"Global Analysis of River Systems: From Earth System Controls to Anthropocene Syndromes" was the title of a 2003 article in the journal *Philosophical Transactions of the Royal Society B*.

"Soils and Sediments in the Anthropocene" ran the headline of a piece from 2004 in the *Journal of Soils and Sediments*.

When Zalasiewicz came across the term, he was intrigued. He noticed that most of those using it were not trained stratigraphers, and he wondered how his colleagues felt about this. At the time, he was head of the stratigraphy committee of the Geological Society of London, the body Lyell and also William Whewell and John Phillips once presided over. At a luncheon meeting, Zalasiewicz asked his fellow committee members what they thought of the Anthropocene. Twenty-one out of the twenty-two thought that the concept had merit.

The group decided to examine the idea as a formal problem in geology. Would the Anthropocene satisfy the criteria used for naming a new epoch? (To geologists, an epoch is a subdivision of a period, which, in turn, is a division of an era: the Holocene, for instance, is an epoch of the Quaternary, which is a period in the Cenozoic.) The answer the members arrived at after a year's worth of study was an unqualified "yes." The sorts of changes that Crutzen had enumerated would, they decided, leave behind "a global stratigraphic signature" that would still be legible millions of years from now, the same way that, say, the Ordovician glaciation left behind a "stratigraphic signature" that is still legible today. Among other things, the members of the group observed in a paper summarizing their findings, the Anthropocene will be marked by a unique "biostratigraphical signal," a product of the current extinction event on the one hand and of the human propensity for redistributing life on the other. This signal will be permanently inscribed, they wrote, "as future evolution will take place from surviving (and frequently anthropogenically relocated) stocks." Or, as Zalasiewicz would have it, rats.

By the time of my visit to Scotland, Zalasiewicz had taken the case for the Anthropocene to the next level. The International

Commission on Stratigraphy, or ICS, is the group responsible for maintaining the official timetable of earth's history. It's the ICS that settles such matters as: when exactly did the Pleistocene begin? (After much heated debate, the commission recently moved that epoch's start date back from 1.8 to 2.6 million years ago.) Zalasiewicz had convinced the ICS to look into formally recognizing the Anthropocene, an effort that, logically enough, he himself was put in charge of. As head of the Anthropocene Working Group, Zalasiewicz is hoping to bring a proposal to a vote by the full body in 2016. If he's successful and the Anthropocene is adopted as a new epoch, every geology textbook in the world immediately will become obsolete.

CHAPTER VI

THE SEA AROUND US

Patella caerulea

Castello Aragonese is a tiny island that rises straight out of the Tyrrhenian Sea, like a turret. Eighteen miles west of Naples, it can be reached from the larger island of Ischia via a long, narrow stone bridge. At the end of the bridge there's a booth where ten euros buys a ticket that allows you to climb—or, better yet, take the elevator—up to the massive castle that gives the island its name. The castle houses a display of medieval torture instruments as well as a fancy hotel and an outdoor café. On a summer evening, the café is supposed to be a pleasant place to sip Campari and contemplate the terrors of the past.

Like many small places, Castello Aragonese is a product of very large forces, in this case the northward drift of Africa, which every year brings Tripoli an inch or so closer to Rome. Along a complicated set of folds, the African plate is pressing into Eurasia, the way a sheet of metal might be forced into a furnace. Occasionally, this process results in violent volcanic eruptions. (One such eruption, in 1302, led the entire population of Ischia to take refuge on Castello Aragonese.) On a more regular basis, it sends streams of gas bubbling out of vents in the sea floor. This gas, as it happens, is almost a hundred percent carbon dioxide.

Carbon dioxide has many interesting properties, one of which is that it dissolves in water to form an acid. I have come to Ischia in late January, deep into the off-season, specifically to swim in its bubbly, acidified bay. Two marine biologists, Jason Hall-Spencer and Maria Cristina Buia, have promised to show me the vents, provided the predicted rainstorm holds off. It is a raw, gray day, and we are thumping along in a fishing boat that's been converted into a research vessel. We round Castello Aragonese and anchor about twenty yards from its rocky cliffs. From the boat, I can't see the vents, but I can see signs of them. A whitish band of barnacles runs all the way around the base of the island, except above the vents, where the barnacles are missing.

"Barnacles are pretty tough," Hall-Spencer observes. He is British, with dirty blond hair that sticks up in unpredictable directions. He's wearing a dry suit, which is a sort of wet suit designed to keep its owner from ever getting wet, and it makes him look as if he's preparing for a space journey. Buia is Italian, with reddish brown hair that reaches her shoulders. She strips down to her bathing suit and pulls on her wet suit with one expert motion. I try to emulate her with a suit I have borrowed for the occasion. It is, I learn as I tug at the zipper, about half a size too small. We all put on masks and flippers and flop in.

The water is frigid. Hall-Spencer is carrying a knife. He pries some sea urchins from a rock and holds them out to me. Their spines are an inky black. We swim on, along the southern shore of the island, toward the vents. Hall-Spencer and Buia keep pausing to gather samples—corals, snails, seaweeds, mussels—which they place in mesh sacs that drag behind them in the water. When we get close enough, I start to see bubbles rising from the sea floor, like beads of quicksilver. Beds of seagrass wave beneath us. The blades are a peculiarly vivid green. This, I later learn, is because the tiny organisms that usually coat them, dulling their color, are missing. The closer we get to the vents, the less there is to collect. The sea urchins drop away, and so, too, do the mussels and the barnacles. Buia finds some hapless limpets attached to the cliff. Their shells have wasted away almost to the point of transparency. Swarms of jellyfish waft by, just a shade paler than the sea.

"Watch out," Hall-Spencer warns. "They sting."

SINCE the start of the industrial revolution, humans have burned through enough fossil fuels—coal, oil, and natural gas—to add some 365 billion metric tons of carbon to the atmosphere. Deforestation has contributed another 180 billion tons. Each year, we throw up another nine billion tons or so, an amount that's been increasing by as much as six percent annually. As a result of all this, the concentration of carbon dioxide in the air today—a little over four hundred parts per million—is higher than at any other point in the last eight hundred thousand years. Quite probably it is higher than at any point in the last several million years. If current trends continue, CO_2 concentrations will top five hundred parts per million, roughly double the levels they were in preindustrial days, by 2050. It is expected that such an increase will produce an eventual average global temperature rise of between three and a half and seven degrees Fahrenheit, and this will, in turn, trigger a variety of world-altering events, including the disappearance of most remaining glaciers, the inundation of low-lying islands and coastal cities, and the melting of the Arctic ice cap. But this is only half the story.

Ocean covers seventy percent of the earth's surface, and everywhere that water and air come into contact there's an exchange. Gases from the atmosphere get absorbed by the ocean and gases dissolved in the ocean are released into the atmosphere. When the two are in equilibrium, roughly the same quantities are being dissolved as are being released. Change the atmosphere's composition, as we have done, and the exchange becomes lopsided: more carbon dioxide enters the water than comes back out. In this way, humans are constantly adding CO_2 to the seas, much as the vents do, but from above rather than below and on a global scale. This year alone the oceans will absorb two and a half billion tons of carbon, and next year it is expected they will absorb another two and a half billion tons. Every day, every American in effect pumps seven pounds of carbon into the sea.

Thanks to all this extra CO_2, the pH of the oceans' surface waters has already dropped, from an average of around 8.2 to an average of around 8.1. Like the Richter scale, the pH scale is logarithmic, so even such a small numerical difference represents a very large real-world change. A decline of .1 means that the oceans are now thirty percent more acidic than they were in 1800. Assuming that humans continue to burn fossil fuels, the oceans will continue to absorb carbon dioxide and will become increasingly acidified. Under what's known as a "business as usual" emissions scenario, surface ocean pH will fall to 8.0 by the middle of this century, and it will drop to 7.8 by the century's end. At that point, the oceans will be 150 percent more acidic than they were at the start of the industrial revolution.*

Owing to the CO_2 pouring out of the vents, the waters around Castello Aragonese provide a near-perfect preview of what lies ahead for the oceans more generally. Which is why I am paddling around the island in January, gradually growing numb from the cold. Here it is possible to swim—even, I think in a moment of panic, to drown—in the seas of tomorrow today.

By the time we get back to the harbor in Ischia, the wind has come up. The deck is a clutter of spent air tanks, dripping wet suits, and chests full of samples. Once unloaded, everything has to be lugged through the narrow streets and up to the local marine biological

*The pH scale runs from zero to fourteen. Seven is neutral; anything above that is basic and below it acidic. Seawater is naturally basic, so as the pH falls the process usually referred to as ocean acidification could, less catchily, be called a decline in ocean alkalinity.

Castello Aragonese.

station, which occupies a steep promontory overlooking the sea. The station was founded by a nineteenth-century German naturalist named Anton Dohrn. Hanging on the wall in the entrance hall, I notice, is a copy of a letter Charles Darwin sent to Dohrn in 1874. In it, Darwin expresses dismay at having heard, through a mutual friend, that Dohrn is overworked.

Installed in tanks in a basement laboratory, the animals Buia and Hall-Spencer gathered from around Castello Aragonese at first appear inert—to my untrained eye, possibly even dead. But after a while, they set about waggling their tentacles and scavenging for food. There is a starfish missing a leg, and a lump of rather rangy-looking coral, and some sea urchins, which move around their tanks on dozens of threadlike "tube feet." (Each tube foot is controlled hydraulically, extending and retracting in response to water pressure.) There is also a six-inch-long sea cucumber, which bears an unfortunate resemblance to a blood sausage or, worse yet, a turd. In the chilly lab, the destructive effect of the vents is plain. *Osilinus turbinatus* is a common Mediterranean snail with a shell of alternating black and white splotches arranged in a snakeskin-like pattern. The *Osilinus turbinatus* in the tank has no pattern; the ridged outer layer of its shell has been eaten away, exposing the smooth, all-white layer underneath. The limpet *Patella caerulea* is shaped like a Chinese straw hat. Several *Patella caerulea* shells have deep lesions through which their owners' putty-colored bodies can be seen. They look as if they have been dunked in acid, which in a manner of speaking they have.

"Because it's so important, we humans put a lot of energy into making sure that the pH of our blood is constant," Hall-Spencer says, raising his voice to be heard over the noise of the running water. "But some of these lower organisms, they don't have the physiology to do that. They've just got to tolerate what's happening outside, and so they get pushed beyond their limits."

Later, over pizza, Hall-Spencer tells me about his first trip to the vents. That was in the summer of 2002, when he was working on an Italian research vessel called the *Urania*. One hot day, the *Urania* was passing by Ischia when the crew decided to anchor and go for a swim. Some of the Italian scientists who knew about the vents took Hall-Spencer to see them, just for the fun of it. He enjoyed the novelty of the experience—swimming through the bubbles is a bit like bathing in champagne—but beyond that, it set him thinking.

At the time, marine biologists were just beginning to recognize the hazards posed by acidification. Some disturbing calculations had been done and some preliminary experiments performed on animals raised in labs. It occurred to Hall-Spencer that the vents could be used for a new and more ambitious sort of study. This one would involve not just a few species reared in tanks, but dozens of species living and breeding in their natural (or, if you prefer, naturally unnatural) environment.

At Castello Aragonese, the vents produce a pH gradient. On the eastern edge of the island, the waters are more or less unaffected. This zone might be thought of as the Mediterranean of the present. As you move closer to the vents, the acidity of the water increases and the pH declines. A map of life along this pH gradient, Hall-Spencer reasoned, would represent a map of what lies ahead for the world's oceans. It would be like having access to an underwater time machine.

It took Hall-Spencer two years to get back to Ischia. He did not yet have funding for his project, and so he had trouble getting anyone to take him seriously. Unable to afford a hotel room, he camped out on a ledge in the cliffs. To collect samples, he used discarded plastic water bottles. "It was a bit Robinson Crusoe-ish," he tells me.

Eventually, he convinced enough people, including Buia, that he was onto something. Their first task was producing a detailed survey of pH levels around the island. Then they organized a census of what was living in each of the different pH zones. This involved placing metal frames along the shore and registering every mussel, barnacle, and limpet clinging to the rocks. It also involved spending hours at a stretch sitting underwater, counting passing fish.

In the waters far from the vents Hall-Spencer and his colleagues found a fairly typical assemblage of Mediterranean species. These included: *Agelas oroides*, a sponge that looks a bit like foam insulation; *Sarpa salpa*, a commonly consumed fish that, on occasion, causes hallucinations; and *Arbacia lixula*, a sea urchin with a lilac tinge. Also living in the area was *Amphiroa rigida*, a spiky, pinkish seaweed, and *Halimeda tuna*, a green seaweed that grows in the shape of a series of connecting disks. (The census was limited to creatures large enough to be seen with the naked eye.) In this vent-free zone, sixty-nine species of animals and fifty-one species of plants were counted.

When Hall-Spencer and his team set up their quadrants closer to the vents, the tally they came up with was very different. *Balanus perforatus* is a grayish barnacle that resembles a tiny volcano. It is common and abundant from west Africa to Wales. In the pH 7.8 zone, which corresponds to the seas of the not-too-distant future, *Balanus perforatus* was gone. *Mytilus galloprovincialis*, a blue-black mussel native to the Mediterranean, is so adaptable that it's established itself in many parts of the world as an invasive. It, too, was missing. Also absent were: *Corallina elongata* and *Corallina officinalis,* both forms of stiff, reddish seaweed; *Pomatoceros triqueter,* a kind of keel worm; three species of coral; several species of snails; and *Arca noae,* a mollusk commonly known as Noah's Ark. All told, one-third of the species found in the vent-free zone were no-shows in the pH 7.8 zone.

"Unfortunately, the biggest tipping point, the one at which the ecosystem starts to crash, is mean pH 7.8, which is what we're expecting to happen by 2100," Hall-Spencer tells me, in his understated British manner. "So that is rather alarming."

SINCE Hall-Spencer's first paper on the vent system appeared, in 2008, there has been an explosion of interest in acidification and its effects. International research projects with names like BIOACID (Biological Impacts of Ocean Acidification) and EPOCA (the European Project on Ocean Acidification) have been funded, and hundreds, perhaps thousands, of experiments have been undertaken. These experiments have been conducted on board ships, in laboratories, and in enclosures known as mesocosms, which allow conditions to be manipulated on a patch of actual ocean.

Again and again, these experiments have confirmed the hazards posed by rising CO_2. While many species will apparently do fine, even thrive in an acidified ocean, lots of others will not. Some of the organisms that have been shown to be vulnerable, like clownfish and Pacific oysters, are familiar from aquariums and the dinner table; others are less charismatic (or tasty) but probably more essential to marine ecosystems. *Emiliania huxleyi,* for example, is a single-celled phytoplankton—a coccolithophore—that surrounds itself with tiny calcite plates. Under magnification, it looks like some kind of crazy crafts project: a soccer ball covered in buttons. It is so common at certain times of year that it turns vast

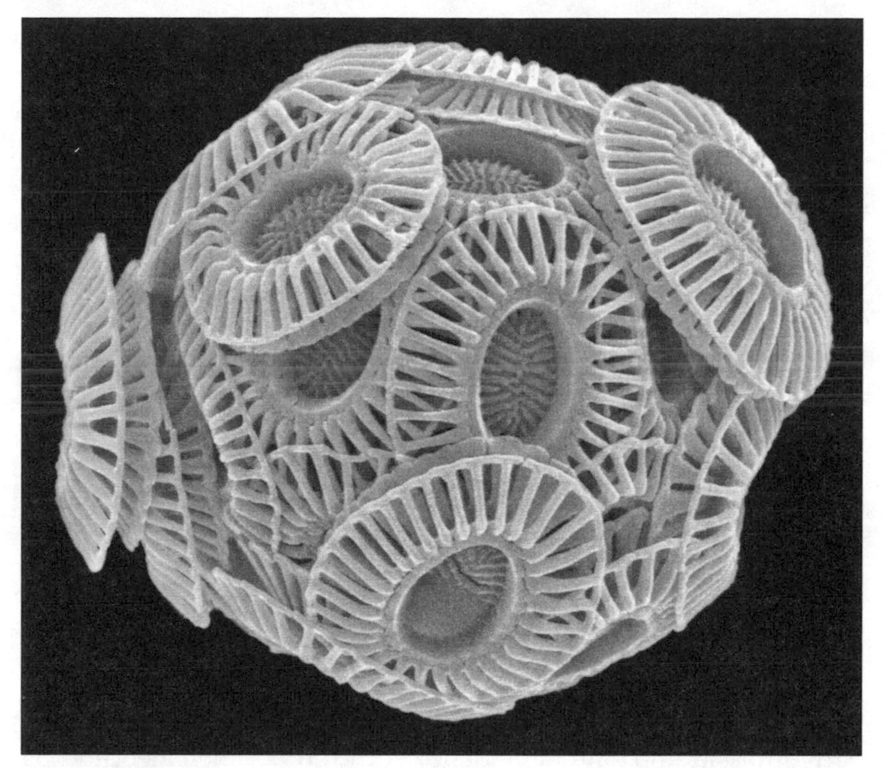

The coccolithophore *Emiliania huxleyi*.

sections of the seas a milky white, and it forms the base of many marine food chains. *Limacina helicina* is a species of pteropod, or "sea butterfly," that resembles a winged snail. It lives in the Arctic and is an important food source for many much larger animals, including herring, salmon, and whales. Both of these species appear to be highly sensitive to acidification: in one mesocosm experiment *Emiliania huxleyi* disappeared altogether from enclosures with elevated CO_2 levels.

Ulf Riebesell is a biological oceanographer at the GEOMAR-Helmholtz Centre for Ocean Research in Kiel, Germany, who has directed several major ocean acidification studies, off the coasts of Norway, Finland, and Svalbard. Riebesell has found that the groups that tend to fare best in acidified water are plankton that are so tiny—less than two microns across—that they form their own microscopic food web. As their numbers increase, these picoplankton, as they are called, use up more nutrients, and larger organisms suffer.

"If you ask me what's going to happen in the future, I think the strongest evidence we have is there is going to be a reduction in biodiversity," Riebesell told me. "Some highly tolerant organisms will become more abundant, but overall diversity will be lost. This is what has happened in all these times of major mass extinction."

Ocean acidification is sometimes referred to as global warming's "equally evil twin." The irony is intentional and fair enough as far as it goes, which may not be far enough. No single mechanism explains all the mass extinctions in the record, and yet changes in ocean chemistry seem to be a pretty good predictor. Ocean acidification played a role in at least two of the Big Five extinctions (the end-Permian and the end-Triassic) and quite possibly it was a major factor in a third (the end-Cretaceous). There's strong evidence for ocean acidification during an extinction event known as the Toarcian Turnover, which occurred 183 million years ago, in the early Jurassic, and similar evidence at the end of the Paleocene, 55 million years ago, when several forms of marine life suffered a major crisis.

"Oh, ocean acidification," Zalasiewicz had told me at Dob's Linn. "That's the big nasty one that's coming down."

WHY is ocean acidification so dangerous? The question is tough to answer only because the list of reasons is so long. Depending on how tightly organisms are able to regulate their internal chemistry, acidification may affect such basic processes as metabolism, enzyme activity, and protein function. Because it will change the makeup of microbial communities, it will alter the availability of key nutrients, like iron and nitrogen. For similar reasons, it will change the amount of light that passes through the water, and for somewhat different reasons, it will alter the way sound propagates. (In general, acidification is expected to make the seas noisier.) It seems likely to promote the growth of toxic algae. It will impact photosynthesis—many plant species are apt to benefit from elevated CO_2 levels—and it will alter the compounds formed by dissolved metals, in some cases in ways that could be poisonous.

Of the myriad possible impacts, probably the most significant involves the group of creatures known as calcifiers. (The term calcifier applies to any organism that builds a shell or external skeleton or, in the case of plants, a kind of internal scaffolding out of the mineral calcium carbonate.) Marine calcifiers are a fantastically varied lot. Echinoderms like starfish and sea urchins are calcifiers, as are mollusks like clams and oysters. So, too, are barnacles, which are crustaceans. Many species of coral are calcifiers; this is how they construct the towering structures that become reefs. Lots of kinds of seaweed are calcifiers; these often feel rigid or brittle to the touch. Coralline algae—minute organisms that grow in colonies that look like a smear of pink paint—are also calcifiers. Brachiopods are calcifiers, and so are coccolithophores, foraminifera, and many types of pteropods—the list goes on and on. It's been estimated that calcification evolved at least two dozen separate times over the course of life's history, and it's quite possible that the number is higher than that.

From a human perspective, calcification looks a bit like construction work and also a bit like alchemy. To build their shells or exoskeletons or calcitic plates, calcifiers must join calcium ions (Ca^{2+}) and carbonate ions (CO_3^{2-}) to form calcium carbonate ($CaCO_3$). But at the concentrations that they're found in ordinary seawater, calcium and carbonate ions won't combine. At the site of calcification, organisms must therefore alter the chemistry of the water to, in effect, impose a chemistry of their own.

Ocean acidification increases the cost of calcification by reducing the number of carbonate ions available to begin with. To extend the construction metaphor, imagine trying to build a house while someone keeps stealing your bricks. The more acidified the water, the greater the energy that's required to complete the necessary steps. At a certain point, the water becomes positively corrosive and solid calcium carbonate begins to dissolve. This is why the limpets that wander too close to the vents at Castello Aragonese end up with holes in their shells.

Lab experiments have indicated that calcifiers will be particularly hard-hit by falling ocean pH, and the list of the disappeared at Castello Aragonese confirms this. In the pH 7.8 zone, three-quarters of the missing species are calcifiers. These include the nearly ubiquitous barnacle *Balanus perforatus,* the hardy mussel *Mytilus galloprovincialis,* and the keel worm *Pomatoceros triqueter.* Other absent calcifiers are *Lima lima,* a common bivalve; *Jujubinus striatus,* a chocolate-colored sea snail; and *Serpulorbis arenarius,* a mollusk known as a worm snail. Calcifying seaweed, meanwhile, is completely absent.

According to geologists who work in the area, the vents at Castello Aragonese have been spewing carbon dioxide for at least several hundred years, maybe longer. Any mussel or barnacle or keel worm that can adapt to lower pH in a time frame of centuries presumably already would have done so. "You give them generations on generations to survive in these conditions, and yet they're not there," Hall-Spencer observed.

And the lower the pH drops, the worse it goes for calcifiers. Right up near the vents, where the bubbles of CO_2 stream up in thick ribbons, Hall-Spencer found that they are entirely absent. In fact, all that remains in this area—the underwater equivalent of a vacant lot—are a few hardy species of native algae, some species of invasive algae, one kind of shrimp, a sponge, and two kinds of sea slugs.

"You won't see *any* calcifying organisms, full stop, in the area where the bubbles are coming up," he told me. "You know how normally in a polluted harbor you've got just a few species that are weedlike and able to cope with massively fluctuating conditions? Well, it's like that when you ramp up CO_2."

* * *

ROUGHLY one-third of the CO_2 that humans have so far pumped into the air has been absorbed by the oceans. This comes to a stunning 150 billion metric tons. As with most aspects of the Anthropocene, though, it's not only the scale of the transfer but also the speed that's significant. A useful (though admittedly imperfect) comparison can be made to alcohol. Just as it makes a big difference to your blood chemistry whether you take a month to go through a six-pack or an hour, it makes a big difference to marine chemistry whether carbon dioxide is added over the course of a million years or a hundred. To the oceans, as to the human liver, rate matters.

If we were adding CO_2 to the air more slowly, geophysical processes, like the weathering of rock, would come into play to counteract acidification. As it is, things are moving too fast for such slow-acting forces to keep up. As Rachel Carson once observed, referring to a very different but at the same time profoundly similar problem: "Time is the essential ingredient, but in the modern world there is no time."

A group of scientists led by Bärbel Hönisch, of Columbia's Lamont-Doherty Earth Observatory, recently reviewed the evidence for changing CO_2 levels in the geologic past and concluded that, although there are several severe episodes of ocean acidification in the record, "no past event perfectly parallels" what is happening right now, owing to "the unprecedented rapidity of CO_2 release currently taking place." It turns out there just aren't many ways to inject billions of tons of carbon into the air very quickly. The best explanation anyone has come up with for the end-Permian extinction is a massive burst of vulcanism in what's now Siberia. But even this spectacular event, which created the formation known as the Siberian Traps, probably released, on an annual basis, less carbon than our cars and factories and power plants.

By burning through coal and oil deposits, humans are putting carbon back into the air that has been sequestered for tens—in most cases hundreds—of millions of years. In the process, we are running geologic history not only in reverse but at warp speed.

"It is the rate of CO_2 release that makes the current great experiment so geologically unusual, and quite probably unprecedented in earth history," Lee Kump, a geologist at Penn State, and Andy Ridgwell, a climate modeler from the University of Bristol, observed in a special issue of the journal *Oceanography* devoted to acidification. Continuing along this path for much longer, the pair continued, "is likely to leave a legacy of the Anthropocene as one of the most notable, if not cataclysmic events in the history of our planet."

CHAPTER VII

DROPPING ACID

Acropora millepora

Half a world away from Castello Aragonese, One Tree Island sits at the southernmost tip of the Great Barrier Reef, about fifty miles off the coast of Australia. It has more than one tree, which surprised me when I got there, expecting—cartoonishly, I suppose—a single palm sticking up out of white sand. As it turned out, there wasn't any sand, either. The whole island consists of pieces of coral rubble, ranging in size from small marbles to huge boulders. Like the living corals they once were part of, the rubble chunks come in dozens of forms. Some are stubby and finger-shaped, others branching, like a candelabra. Still others resemble antlers or dinner plates or bits of brain. It is believed that One Tree Island was created during a particularly vicious storm that occurred some four thousand years ago. (As one geologist who has studied the place put it to me, "You wouldn't have wanted to be there when that happened.") The island is still in the process of changing shape; a storm that passed through in March 2009—Cyclone Hamish—added a ridge that runs along the island's eastern shore.

One Tree would qualify as deserted except for a tiny research station operated by the University of Sydney. I traveled to the island, as just about everyone does, from another, slightly larger island about twelve miles away. (That island is known as Heron Island, also a misnomer, since at Heron there are no herons.) When we docked—or really moored, since One Tree has no dock—a loggerhead turtle was heaving herself out of the water onto the shore. She was nearly four feet long, with a large welt on her shell, which was encrusted with ancient-looking barnacles. News travels fast on a nearly deserted island, and soon the entire human population of One Tree—twelve people, including me—had come out to watch. Sea turtles usually lay their eggs at night, on sandy beaches; this was in the middle of the day, on jagged coral rubble. The turtle tried to dig a hole with her back flippers. After much exertion, she produced a shallow trough. By this point, one of her flippers was bleeding. She heaved herself farther up the shore and tried again, with similar results. She was still at it an hour and a half later, when I had to go get a safety lecture from the manager of the research station, Russell

One Tree Island and its surrounding reef, as seen from the air.

Graham. He warned me not to go swimming when the tide was going out, as I might find myself "swept off to Fiji." (This was a line I would hear repeated many times during my stay, though there was some disagreement about whether the current was heading toward Fiji or really away from it.) Once I'd taken in this and other advisories—the bite of a blue-ringed octopus is usually fatal; the sting of a stonefish is not, but it is so painful it will make you wish it were—I went back to see how the turtle was doing. Apparently, she had given up and crawled back into the sea.

The One Tree Island Research Station is a bare-bones affair. It consists of two makeshift labs, a pair of cabins, and an outhouse with a composting toilet. The cabins rest directly on the rubble, for the most part with no floor, so that even when you're indoors you feel as if you're out. Teams of scientists from all around the world book themselves into the station for stays of a few weeks or a few months. At one point, someone must have decided that every team should leave a record of its visit on the cabin walls. GETTING TO THE CORE IN 2004, reads one inscription, drawn in magic marker. Others include:

THE CRAB CREW: CLAWS FOR A CAUSE—2005
CORAL SEX—2008
THE FLUORESCENCE TEAM—2009

The American-Israeli team that was in residence at the time of my arrival had already made two trips to the island. The epigram from its first visit, DROPPING ACID ON CORALS, was accompanied by a sketch of a syringe dripping what looked like blood onto a globe. The group's latest message referred to its study site, a patch of coral known as DK-13. DK-13 lies out on the reef, far enough away from the station that, for the purposes of communication, it might as well be on the moon.

The writing on the wall said, DK-13: NO ONE CAN HEAR YOU SCREAM.

* * *

THE first European to encounter the Great Barrier Reef was Captain James Cook. In the spring of 1770, Cook was sailing along the east coast of Australia when his ship, the *Endeavour*, rammed into a section of the reef about thirty miles southeast of what is now, not coincidentally, Cooktown. Everything dispensable, including the ship's cannon, was tossed overboard, and the leaky *Endeavour* managed to creak ashore, where the crew spent the next two months repairing its hull. Cook was flummoxed by what he described as "a wall of Coral Rock rising all most perpendicular out of the unfathomable Ocean." He understood that the reef was biological in origin, that it had been "formed in the Sea by animals." But how, then, he would later ask, had it come to be "thrown up to such a height?"

The question of how coral reefs arose was still an open one sixty years later, when Lyell sat down to write the *Principles*. Although he had never seen a reef, Lyell was fascinated by them, and he devoted part of volume two to speculating about their origins. Lyell's theory—that reefs grew from the rims of extinct underwater volcanoes—he borrowed more or less wholesale from a Russian naturalist named Johann Friedrich von Eschscholtz. (Before Bikini Atoll became Bikini Atoll, it was called, rather less enticingly, Eschsholtz Atoll.)

When his turn came to theorize about reefs, Darwin had the advantage of actually having visited some. In November 1835, the *Beagle* moored off Tahiti. Darwin climbed to one of the highest points on the island, and from there he could survey the neighboring island of Moorea. Moorea, he observed, was encircled by a reef the way a framed etching is surrounded by a mat.

"I am glad that we have visited these islands," Darwin wrote in his diary, for coral reefs "rank high amongst the wonderful objects in the world." Looking over at Moorea and its surrounding reef, he pictured time running forward; if the island were to sink away,

Moorea's reef would become an atoll. When Darwin returned to London and shared his subsidence theory with Lyell, Lyell, though impressed, foresaw resistance. "Do not flatter yourself that you will be believed until you are growing bald like me," he warned.

In fact, debate about Darwin's theory—the subject of his 1842 book *The Structure and Distribution of Coral Reefs*—continued until the nineteen-fifties, when the U.S. Navy arrived in the Marshall Islands with plans to vaporize some of them. In preparation for the H-bomb tests, the Navy drilled a series of cores on an atoll called Enewetak. As one of Darwin's biographers put it, these cores proved his theory to be, in its large lines at least, "astoundingly correct."

Darwin's description of coral reefs as "amongst the wonderful objects of the world" also still stands. Indeed, the more that has been learned about reefs, the more marvelous they seem. Reefs are organic paradoxes—obdurate, ship-destroying ramparts constructed

Coral polyps.

by tiny gelatinous creatures. They are part animal, part vegetable, and part mineral, at once teeming with life and, at the same time, mostly dead.

Like sea urchins and starfish and clams and oysters and barnacles, reef-building corals have mastered the alchemy of calcification. What sets them apart from other calcifiers is that instead of working solo, to produce a shell, say, or some calcitic plates, corals engage in vast communal building projects that stretch over generations. Each individual, known unflatteringly as a polyp, adds to its colony's collective exoskeleton. On a reef, billions of polyps belonging to as many as a hundred different species are all devoting themselves to this same basic task. Given enough time (and the right conditions), the result is another paradox: a living structure. The Great Barrier Reef extends, discontinuously, for more than fifteen hundred miles, and in some places it is five hundred feet thick. By the scale of reefs, the pyramids at Giza are kiddie blocks.

The way corals change the world—with huge construction projects spanning multiple generations—might be likened to the way that humans do, with this crucial difference. Instead of displacing other creatures, corals support them. Thousands—perhaps millions—of species have evolved to rely on coral reefs, either directly for protection or food, or indirectly, to prey on those species that come seeking protection or food. This coevolutionary venture has been under way for many geologic epochs. Researchers now believe it won't last out the Anthropocene. "It is likely that reefs will be the first major ecosystem in the modern era to become ecologically extinct" is how a trio of British scientists recently put it. Some give reefs until the end of the century, others less time even than that. A paper published in *Nature* by the former head of the One Tree Island Research Station, Ove Hoegh-Guldberg, predicted that if current trends continue, then by around 2050 visitors to the Great Barrier Reef will arrive to find "rapidly eroding rubble banks."

* * *

I CAME to One Tree more or less by accident. My original plan had been to stay on Heron Island, where there's a much larger research station and also a ritzy resort. On Heron, I was going to watch the annual coral spawning and observe what had been described to me in various Skype conversations as a seminal experiment on ocean acidification. Researchers from the University of Queensland were building an elaborate Plexiglas mesocosm that was going to allow them to manipulate CO_2 levels on a patch of reef, even as it allowed the various creatures that depend on the reef to swim in and out. By changing the pH inside the mesocosm and measuring what happened to the corals, they were going to be able to generate predictions about the reef as a whole. I arrived at Heron in time to see the spawning—more on this later—but the experiment was way behind schedule and the mesocosm still in pieces. Instead of the reef of the future, all there was to see was a bunch of anxious graduate students hunched over soldering irons in the lab.

As I was trying to figure out what to do next, I heard about another experiment on corals and ocean acidification that was under way at One Tree, which, by the scale of the Great Barrier Reef, lies just around the corner. Three days later—there is no regular transportation to One Tree—I managed to get a boat over.

The head of the team at One Tree was an atmospheric scientist named Ken Caldeira. Caldeira, who's based at Stanford, is often credited with having coined the term "ocean acidification." He became interested in the subject in the late nineteen-nineties when he was hired to do a project for the Department of Energy. The department wanted to know what the consequences would be of capturing carbon dioxide from smokestacks and injecting it into the deep sea. At that point, almost no modeling work had been done on the effects of carbon emissions on the oceans. Caldeira set about calculating how the ocean's pH would change as a result of deep-sea injection, and then compared that result with the current practice of pumping CO_2 into the atmosphere and allowing it to be absorbed by surface waters. In 2003, he submitted his results to *Nature*. The journal's editors advised him to drop the discussion of deep-ocean injection because the calculations concerning the effects of ordinary atmospheric release were so startling. Caldeira published the first part of his paper under the subheading "The Coming Centuries May See More Ocean Acidification Than the Past 300 Million Years."

"Under business as usual, by mid-century things are looking rather grim," he told me a few hours after I had arrived at One Tree. We were sitting at a beat-up picnic table, looking out over the heartbreaking blue of the Coral Sea. The island's large and boisterous population of terns was screaming in the background. Caldeira paused: "I mean, they're looking grim already."

CALDEIRA, who is in his mid-fifties, has curly brown hair, a boyish smile, and a voice that tends to rise toward the end of sentences, so that it often seems he is posing a question even when he's not. Before getting into research, he worked as a software developer on Wall Street. One of his clients was the New York Stock Exchange, for whom he designed a computer program to detect insider trading. The program functioned as it was supposed to, but after a while Caldeira came to believe that the NYSE wasn't really interested in catching insider traders, and he decided to switch professions.

Unlike most atmospheric scientists, who focus on one particular aspect of the system, Caldeira is, at any given moment, working on four or five disparate projects. He particularly likes computations of a provocative or surprising nature; for example, he once calculated that cutting down all the world's forests and replacing them with grasslands would have a slight cooling effect. (Grasslands, which are lighter in color than forests, absorb less sunlight.) Other calculations of his show that to keep pace with the present rate of temperature change, plants and animals would have to migrate poleward by thirty feet a day, and that a molecule of CO_2 generated by burning fossil fuels will, in the course of its lifetime in the atmosphere, trap a hundred thousand times more heat than was released in producing it.

At One Tree, life for Caldeira and his team revolved around the tides. An hour before the first low tide of the day and then an hour afterward, someone had to collect water samples out at DK-13, so named because the Australian researcher who had set up the site, Donald Kinsey, had labeled it with his initials. A little more than twelve hours later, the process would be repeated, and so on, from one low tide to the next. The experiment was slow tech rather than high tech; the idea was to measure various properties of the water that Kinsey had measured back in the nineteen-seventies, then compare the two sets of data and try to tease out how calcification rates on the reef had changed in the intervening decades. In daylight, the trip to DK-13 could be made by one person. In the dark, in deference to the fact that "no one can hear you scream," the rule was that two had to go.

My first evening on One Tree, low tide fell at 8:53 PM. Caldeira was making the post–low-tide trip, and I volunteered to go with him. At around nine o'clock, we gathered up half a dozen sampling bottles, a pair of flashlights, and a handheld GPS unit and started out.

From the research station, it was about a mile walk to DK-13. The route, which someone had plugged into the GPS unit, led around the southern tip of the island and over a slick expanse of rubble that had been nicknamed the "algal highway." From there it veered out onto the reef itself.

Since corals like light but can't survive long exposure to the air, they tend to grow as high as the water level at low tide and then spread out laterally. This produces an expanse of reef that's more or less flat, like a series of tables, which can be crossed the way a kid, after school, might jump from desk to desk. The surface of One Tree's reef flat was brittle and brownish and was known around the research station as the "pie crust." It crackled ominously underfoot. Caldeira warned me that if I fell through, it would be bad for the reef and even worse for my shins. I recalled another message I had seen penned on the wall of the research station: DON'T TRUST THE PIE CRUST.

The night was balmy and, beyond the beams of our flashlights, pitch-black. Even in the dark, the extraordinary vitality of the reef was evident. We passed several loggerhead turtles waiting out low tide with what looked like bored expressions. We encountered bright blue starfish, and leopard sharks stranded in shallow pools, and ruddy octopuses doing their best to blend into the reef. Every few feet, we had to step over a giant clam, which appeared to be leering with garishly painted lips. (The mantles of giant clams are packed with colorful symbiotic algae.) The sandy strips between the blocks of coral were littered with sea cucumbers, which, despite the name, are animals whose closest relations are sea urchins. On the Great Barrier Reef, the sea cucumbers are the size not of cucumbers but of bolster cushions. Out of curiosity, I decided to pick one up. It was about two feet long and inky black. It felt like slime-covered velvet.

After a few wrong turns and several delays while Caldeira tried to photograph the octopuses with a waterproof camera, we reached DK-13. The site consisted of nothing more than a yellow buoy and some sensing equipment anchored to the reef with a rope. I glanced back in what I thought was the direction of the island, but there was no island, or land of any sort, to be seen. We rinsed out the sampling bottles, filled them, and started back. The darkness was, if anything, even more complete. The stars were so bright they appeared to be straining out of the sky. For a brief moment I felt I understood what it must have been like for an explorer like Cook to arrive at such a place, at the edge of the known world.

CORAL reefs grow in a great swath that stretches like a belt around the belly of the earth, from thirty degrees north to thirty degrees south latitude. After the Great Barrier Reef, the world's second-largest reef is off the coast of Belize. There are extensive coral reefs in the tropical Pacific, in the Indian Ocean, and in the Red Sea, and many smaller ones in the Caribbean. Yet curiously enough, the first evidence that CO_2 could kill a reef came from Arizona, from the self-enclosed, supposedly self-sufficient world known as Biosphere 2.

A three-acre, glassed-in structure shaped like a ziggurat, Biosphere 2 was built in the late nineteen-eighties by a private group largely funded by the billionaire Edward Bass. It was intended to demonstrate how life on earth—Biosphere 1—could be re-created on, say, Mars. The building contained a "rainforest," a "desert," an "agricultural zone," and an artificial "ocean." The first group of Biospherians, four men and four women, remained sealed inside the place for two years. They grew all of their own food and, for a stretch, breathed only recycled air. Still, the project was widely considered a failure. The Biospherians spent much of their time hungry, and, even more ominously, they lost control of their artificial atmosphere. In the various "ecosystems," decomposition, which takes up oxygen and gives off carbon dioxide, was supposed to be balanced by photosynthesis, which does the reverse. For reasons mainly having to do with the richness of the soil that had been imported into the "agricultural zone," decomposition won out. Oxygen levels inside the building fell sharply, and the Biospherians developed what amounted to altitude sickness. Carbon dioxide levels, meanwhile, soared. Eventually, they reached three thousand parts per million, roughly eight times the levels outside.

Biosphere 2 officially collapsed in 1995, and Columbia University took over the management of the building. The "ocean," a tank the size of an Olympic swimming pool, was by this point a wreck: most of the fish it had been stocked with were dead, and the corals were just barely hanging on. A marine biologist named Chris Langdon was assigned the task of figuring out something educational to do with the tank. His first step was to adjust the water chemistry.

Not surprisingly, given the high CO_2 content of the air, the pH of the "ocean" was low. Langdon tried to fix this, but strange things kept happening. Figuring out why became something of an obsession. After a while, Langdon sold his house in New York and moved to Arizona, so that he could experiment on the "ocean" full-time.

Although the effects of acidification are generally expressed in terms of pH, there's another way to look at what's going on that's just as important—to many organisms probably more important—and this is in terms of a property of seawater known, rather cumbersomely, as the "saturation state with respect to calcium carbonate," or, alternatively, the "saturation state with respect to aragonite." (Calcium carbonate comes in two different forms, depending on its crystal structure; aragonite, which is the form corals manufacture, is the more soluble variety.) The saturation state is determined by a complicated chemical formula; essentially, it's a measure of the concentration of calcium and carbonate ions floating around. When CO_2 dissolves in water, it forms carbonic acid—H_2CO_3—which effectively "eats" carbonate ions, thus lowering the saturation state.

When Langdon showed up at Biosphere 2, the prevailing view among marine biologists was that corals did not much care about the saturation state as long as it remained above one. (Below one, water is "undersaturated," and calcium carbonate dissolves.) Based on what he was seeing, Langdon became convinced that corals *did* care about the saturation state; indeed, they cared about it deeply. To test his hypothesis, Langdon employed a straightforward, if time-consuming, procedure. Conditions in the "ocean" would be varied, and small colonies of corals, which were attached to little tiles, would be periodically lifted out of the water and weighed. If the colony was putting on weight, it would show that it was growing—adding more mass through calcification. The experiment took more than three years to complete and yielded more than a thousand measurements. It revealed a more or less linear relationship between the growth rate of the corals and the saturation state of the water. Corals grew fastest at an aragonite saturation state of five, slower at four, and still slower at three. At a level of two, they basically quit building, like frustrated contractors throwing up their hands. In the artificial world of Biosphere 2, the implications of this discovery were interesting. In the real world—Biosphere 1—they were rather more worrisome.

Prior to the industrial revolution, all of the world's major reefs could be found in water with an aragonite saturation state between four and five. Today, there's almost no place left on the planet where the saturation state is above four, and if current emissions trends continue, by 2060 there will be no regions left above 3.5. By 2100, none will remain above three. As saturation levels fall, the energy required for calcification will increase, and calcification rates will decline. Eventually, saturation levels may drop so low that corals quit calcifying altogether, but long before that point, they will be in trouble. This is because out in the real world, reefs are constantly being eaten away at by fish and sea urchins and burrowing worms. They are also being battered by waves and storms, like the one that created One Tree. Thus, just to hold their own, reefs must always be growing.

"It's like a tree with bugs," Langdon once told me. "It needs to grow pretty quickly just to stay even."

Langdon published his results in 2000. At that point many marine biologists were skeptical, in no small part, it seems, because of his association with the discredited Biosphere project. Langdon spent another two years redoing his experiments, this time with even tighter controls. The findings were the same. In the meantime, other researchers launched their own studies. These, too, confirmed Langdon's discovery: reef-building corals are sensitive to the saturation state. This has now been shown in dozens more lab studies and also on an actual reef. A few years ago, Langdon and some colleagues conducted an experiment on a stretch of reef near a volcanic vent system off Papua New Guinea. The experiment, modeled on Hall-Spencer's work at Castello Aragonese, again used the volcanic vents as a natural source of acidification. As the saturation state of the water dropped, coral diversity plunged. Coralline algae declined even more drastically, an ominous sign since coralline algae act like a kind of reef glue, cementing the structure together. Seagrass, meanwhile, thrived.

"A few decades ago I, myself, would have thought it ridiculous to imagine that reefs might have a limited lifespan," J. E. N. Veron, former chief scientist of the Australian Institute of Marine Science, has written. "Yet here I am today, humbled to have spent the most productive scientific years of my life around the rich wonders of the underwater world, and utterly convinced that they will not be there for our children's children to enjoy." A recent study by a team of Australian researchers found that coral cover in the Great Barrier Reef has declined by fifty percent just in the last thirty years.

Not long before their trip to One Tree, Caldeira and some of the other members of his team published a paper assessing the future of corals, using both computer models and data gathered in the field. The paper concluded that if current emissions trends continue, within the next fifty years or so "all coral reefs will cease to grow and start to dissolve."

IN between trips out over the reef to collect samples, the scientists at One Tree did a lot of snorkeling. The group's preferred spot was about a half a mile offshore, on the opposite side of the island from DK-13, and getting there meant cajoling Graham, the station manager, into taking out the boat, something that he did only with reluctance and a fair amount of grumbling.

Some of the scientists, who had dived all over—in the Philippines, in Indonesia, in the Caribbean, and in the South Pacific—told me that the snorkeling at One Tree was about as good as it gets. I found this easy to believe. The first time I jumped off the boat and looked down at the swirl of life beneath me, it felt unreal, as if I'd swum into the undersea world of Jacques Cousteau. Schools of small fish were followed by schools of larger fish, which were followed by sharks. Huge rays glided by, trailed by turtles the size of bathtubs. I tried to keep a mental list of what I'd seen, but it was like trying to catalog a dream. After each outing, I spent hours looking through a huge volume called *The Fishes of the Great Barrier Reef and the Coral Sea*. Among the fish that I think I may have spotted were: tiger sharks, lemon sharks, gray reef sharks, blue-spine unicorn fish, yellow boxfish, spotted boxfish, conspicuous angelfish, Barrier Reef anemonefish, Barrier Reef chromis, minifin parrotfish, Pacific longnose parrotfish, somber sweetlips, fourspot herring, yellowfin tuna, common dolphinfish, deceiver fangblenny, yellow spotted sawtail, barred rabbitfish, blunt-headed wrasse, and striped cleaner wrasse.

Reefs are often compared to rainforests, and in terms of the sheer variety of life, the comparison is apt. Choose just about any group you like, and the numbers are staggering. An Australian researcher once broke apart a volleyball-sized chunk of coral and found, living inside of it, more than fourteen hundred polychaete worms belonging to 103 different species. More recently, American researchers cracked open chunks of corals to look for crustaceans; in a square meter's worth collected near Heron Island, they found representatives of more than a hundred species, and in a similar-sized sample, collected at the northern tip of the Great Barrier Reef, they found representatives of more than a hundred and twenty. It is estimated that at least half a million and possibly as many as nine million species spend as least part of their lives on coral reefs.

This diversity is all the more astonishing in light of the underlying conditions. Tropical waters tend to be low in nutrients, like nitrogen and phosphorus, which are crucial to most forms of life. (This has to do with what's called the thermal structure of the water column, and it's why tropical waters are often so beautifully clear.) As a consequence, the seas in the tropics should be barren—the aqueous equivalent of deserts. Reefs are thus not just underwater rainforests; they are rainforests in a marine Sahara. The first person to be perplexed by this incongruity was Darwin, and it has since become known as "Darwin's paradox." Darwin's paradox has never been entirely resolved, but one key to the puzzle seems to be recycling. Reefs—or, really, reef creatures—have developed a fantastically efficient system by which nutrients are passed from one class of organisms to another, as at a giant bazaar. Corals are the main players in this complex system of exchange, and, at the same time, they provide the platform that makes the trading possible. Without them, there's just more watery desert.

"Corals build the architecture of the ecosystem," Caldeira told me. "So it's pretty clear if they go, the whole ecosystem goes."

One of the Israeli scientists, Jack Silverman, put it to me this way: "If you don't have a building, where are the tenants going to go?"

REEFS have come and gone several times in the past, and their remains crop up in all sorts of unlikely places. The ruins of reefs from the Triassic, for example, can now be found towering thousands of feet above sea level in the Austrian Alps. The Guadalupe Mountains in west Texas are what's left of reefs from the Permian period that were elevated in an episode of "tectonic compression" about eighty million years ago. Reefs from the Silurian period can be seen in northern Greenland.

All these ancient reefs consist of limestone, but the creatures that created them were quite different. Among the organisms that built reefs in the Cretaceous were enormous bivalves known as rudists. In the Silurian, reef builders included spongelike creatures called stromatoporoids, or "stroms" for short. In the Devonian, reefs were constructed by rugose corals, which grew in the shape of horns, and tabulate corals, which grew in the shape of honeycombs. Both rugose corals and tabulate corals were only distantly related to today's scleractinian corals, and both orders died out in the great extinction at the end of the Permian. This extinction shows up in the geologic record as (among other things) a "reef gap"—a period of about ten million years when reefs went missing altogether. Reef gaps also occurred after the late Devonian and the late Triassic extinctions, and in each of these cases it also took millions of years for reef construction to resume. This correlation has prompted some scientists to argue that reef building as an enterprise must be particularly vulnerable to environmental change—yet another paradox, since reef building is also one of the oldest enterprises on earth.

Ocean acidification is, of course, not the only threat reefs are under. Indeed, in some parts of the world, reefs probably will not last long enough for ocean acidification to finish them off. The roster of perils includes, but is not limited to: overfishing, which promotes the growth of algae that compete with corals; agricultural runoff, which also encourages algae growth; deforestation, which leads to siltation and reduces water clarity; and dynamite fishing, whose destructive potential would seem to be self-explanatory. All of these stresses make corals susceptible to pathogens. White-band disease is a bacterial infection that, as the name suggests, produces a band of white necrotic tissue. It afflicts two species of Caribbean coral, *Acropora palmata* (commonly known as elkhorn coral) and *Acropora cervicornis* (staghorn coral), which until recently were the dominant reef builders in the region. The disease has so ravaged the two species that both are now listed as "critically endangered" by the International Union for Conservation of Nature. Meanwhile coral cover in the Caribbean has in recent decades declined by close to eighty percent.

Finally and perhaps most significant on the list of hazards is climate change—ocean acidification's equally evil twin.

Tropical reefs need warmth, but when water temperatures rise too high, trouble ensues. The reasons for this have to do with the fact that reef-building corals lead double lives. Each individual polyp is an animal and, at the same time, a host for microscopic plants known as zooxanthellae. The zooxanthellae produce carbohydrates, via photosynthesis, and the polyps harvest these carbohydrates, much as farmers harvest corn. Once water temperatures rise past a certain point—that temperature varies by location and also by species—the symbiotic relation between the corals and their tenants breaks down. The zooxanthellae begin to produce dangerous concentrations of oxygen radicals, and the polyps respond, desperately and often self-defeatingly, by expelling them. Without the zooxanthellae, which are the source of their fantastic colors, the corals appear to turn white—this is the phenomenon that's become known as "coral bleaching." Bleached colonies stop growing and, if the damage is severe enough, die. There were major bleaching events in 1998, 2005, and 2010, and the frequency and intensity of such events are expected to increase as global temperatures climb. A study of more than eight hundred reef-building coral species, published in *Science* in 2008, found a third of them to be in danger of extinction, largely as a result of rising ocean temperatures. This has made stony corals one of the most endangered groups on the planet: the proportion of coral species ranked as "threatened," the study noted, exceeds "that of most terrestrial animal groups apart from amphibians."

ISLANDS are worlds in miniature or, as the writer David Quammen observed, "almost a caricature of nature's full complexity." By this account, One Tree is a caricature of a caricature. The whole place is less than 750 feet long and 500 feet wide, yet hundreds of scientists have worked there, drawn to it, in many cases, by its very diminutiveness. In the nineteen-seventies, a trio of Australian scientists set about producing a complete biological census of the island. They spent the better part of three years living in tents and cataloging every single plant and animal species they could find, including: trees (3 species), grasses (4 species), birds (29 species), flies (90 species) and mites (102 species). The island, they discovered, has no resident mammals, unless you count the scientists themselves or a pig that was once brought over and kept in a cage until it was barbecued. The monograph that resulted from this research ran to four hundred pages. It opened with a poem attesting to the charms of the tiny cay:

> An island slumbering—
> Clasped in a shimmering circlet
> Of waters turquoise and blue.
> Guarding her jewel from the pounding surf
> On her coral rim.

On my last day at One Tree, no snorkeling trips were planned, so I decided to try to walk across the island, an exercise that should have taken about fifteen minutes. Not very far into my journey, I ran into Graham, the station manager. A rangy man with bright blue eyes, ginger-colored hair, and a walrus mustache, Graham looked to me like he would have made an excellent pirate. We fell into walking and talking together, and as we wandered along, Graham kept picking up bits of plastic that the waves had carried to One Tree: the cap of a bottle; a scrap of insulation, probably from a ship's door; a stretch of PVC pipe. He had a whole collection of these bits of flotsam, which he displayed in a wire cage; the point of the exhibit, he told me, was to demonstrate to visitors "what our race is doing."

Graham offered to show me how the research station actually functioned, and so we threaded our way behind the cabins and the labs, toward the island's midsection. It was breeding season, and everywhere we walked, there were birds strutting around, screaming: bridled terns, which are black on top and white on their chests; lesser crested terns, which are gray with black and white faces; and black noddies, which have a patch of white on their heads. I could see why humans had had such an easy time killing off nesting seabirds; the terns seemed completely unafraid and were so much underfoot it took an effort not to step on them.

Graham brought me to see the photovoltaic panels that provide the research station with power, and the tanks for collecting rain to supply it with water. The tanks were mounted on a platform, and from it we could look over the tops of the island's trees. According to my very rough calculations, these numbered around five hundred. They seemed to be growing directly out of the rubble, like flagpoles. Just beyond the edge of the platform, Graham pointed out a bridled tern that was pecking at a black noddy chick. Soon, the chick was dead. "She won't eat him," he predicted, and he was right. The bridled tern walked away from the chick, who shortly thereafter was consumed by a gull. Graham was philosophical about the episode, versions of which he had obviously seen many times; it would keep the island's bird population from outstripping its resources.

That night was the first night of Hanukkah. For the holiday, someone had crafted a menorah out of a tree branch and strapped two candles onto it with duct tape. Lighted out on the beach, the makeshift menorah sent shadows skittering across the rubble. Dinner that evening was kangaroo meat, which I found surprisingly tasty, but which, the Israelis noted, was distinctly not kosher.

Later, I set out for DK-13 with a postdoc named Kenny Schneider. By this point, the tides had crept forward by more than two hours, so Schneider and I were scheduled to arrive at the site a few minutes before midnight. Schneider had made the journey before but still hadn't quite mastered the workings of the GPS unit. About halfway there, we found that we had wandered off the prescribed route. The water was soon up to our chests. This made walking that much slower and more difficult, and the tide was now coming in. A variety of anxious thoughts ran through my mind. Would we be able to swim back to the station? Would we even be able to figure out the right direction to swim in? Would we finally settle the Fiji question?

Long after we were supposed to, Schneider and I spotted the yellow buoy of DK-13. We filled the sampling bottles and headed back. I was struck again by the extraordinary stars and the lightless horizon. I also felt, as I had several times at One Tree, the incongruity of my position. The reason I'd come to the Great Barrier Reef was to write about the scale of human influence. And yet Schneider and I seemed very, very small in the unbroken dark.

LIKE the Jews, the corals of the Great Barrier Reef observe a lunar calendar. Once a year, after a full moon at the start of the austral summer, they engage in what's known as mass spawning—a kind of synchronized group sex. I was told that the mass spawning was a spectacle not to be missed, and so I planned my trip to Australia accordingly.

For the most part, corals are extremely chaste; they reproduce asexually, by "budding." The annual spawning is thus a rare opportunity to, genetically speaking, mix things up. Most spawners are hermaphrodites, meaning that a single polyp produces both eggs and sperm, all wrapped together in a convenient little bundle. No one knows exactly how corals synchronize their spawning, but they are believed to respond to both light and temperature.

In the buildup to the big night—the mass spawning always occurs after sundown—the corals begin to "set," which might be thought of as the scleractinian version of going into labor. The egg-sperm bundles start to bulge out from the polyps, and the whole colony develops what looks like goose bumps. Back on Heron Island, some Australian researchers had set up an elaborate nursery so they could study the event. They had gathered up colonies of some of the most common species on the reef, including *Acropora millepora*, which, as one of the scientists put it to me, functions

Acropora millepora in the process of spawning.

as the "lab rat" of the coral world, and were raising them in tanks. *Acropora millepora* produces a colony that looks like a cluster of tiny Christmas trees. No one was allowed to go near the tanks with a flashlight, for fear that it would upset the corals' internal clocks. Instead everyone was wearing special red headlamps. With a borrowed headlamp, I could see the egg-sperm bundles straining against the polyps' transparent tissue. The bundles were pink and resembled glass beads.

The head of the team, a researcher named Selina Ward, from the University of Queensland, bustled around the tanks of gravid corals like an obstetrician preparing for a delivery. She told me that each bundle held somewhere between twenty and forty eggs and probably thousands of sperm. Not long after they were released, the bundles would break open and spill their gametes, which, if they managed to find partners, would result in tiny pink larvae. As soon as the corals in her tanks spawned, Ward was planning to scoop up the bundles and subject them to different levels of acidification. She had been studying the effects of acidification on spawning for the past several years, and her results suggested that lower saturation levels led to significant declines in fertilization. Saturation levels also affected larval development and settlement—the process by which coral larvae drop out of the water column, attach themselves to something solid, and start producing new colonies.

"Broadly speaking, all our results have been negative so far," Ward told me. "If we continue the way we are, without making dramatic changes to our carbon emissions immediately, I think we're looking at a situation where, in the future, what we've got at best is remnant patches of corals."

Later that night, some of the other researchers at Heron Island, including the graduate students who were trying to weld together the overdue mesocosm, heard that Ward's corals were getting ready to spawn and organized a nocturnal snorkel. This was a much more elaborate affair than the snorkeling trips at One Tree, complete with wet suits and underwater lights. There wasn't enough equipment for everyone to go at once, so we went in two shifts. I was in the first, and initially I was disappointed, because nothing seemed to be happening. Then, after a while, I noticed a few corals releasing their bundles. Almost immediately, countless others followed. The scene resembled a blizzard in the Alps, only in reverse. The water filled with streams of pink beads floating toward the surface, like snow falling upward. Iridescent worms appeared to eat the bundles, producing an eerie glow, and a slick of mauve began to form on the surface. When my shift was over, I reluctantly climbed out of the water and handed over my light.

CHAPTER VIII

THE FOREST AND THE TREES

Alzatea verticillata

"TREES ARE STUNNING," MILES SILMAN WAS SAYING. "THEY ARE very beautiful. It's true they take a little more appreciation. You walk into a forest, and the first thing you notice is, 'That's a big tree,' or 'That's a tall tree,' but when you start to think about their life history, about everything that goes into getting a tree to that spot, it's really neat. It's kind of like wine; once you start to understand it, it becomes more intriguing." We were standing in eastern Peru, at the edge of the Andes, on top of a twelve-thousand-foot-high mountain, where, in fact, there were no trees—just scrub and, somewhat incongruously, a dozen or so cows, eyeing us suspiciously. The sun was sinking, and with it the temperature, but the view, in the orange glow of evening, was extraordinary. To the east was the ribbon of the Alto Madre de Dios River, which flows into the Beni River, which flows into the Madeira River, which eventually meets the Amazon. Spread out before us was Manú National Park, one of the world's great biodiversity "hot spots."

"In your field of vision is one out of every nine bird species on the planet," Silman told me. "Just in our plots alone, we have over a thousand species of trees."

Silman and I and several of Silman's Peruvian graduate students had just arrived on the mountaintop, having set out that morning from the city of Cuzco. As the crow flies, the distance we'd traveled was only about fifty miles, but the trip had taken us an entire day of driving along serpentine dirt roads. The roads wound past villages made of mud brick and fields perched at improbable angles and women in colorful skirts and brown felt hats carrying babies in slings on their backs. At the largest of the towns, we'd stopped to have lunch and purchase provisions for a four-day hike. These included bread and cheese and a shopping bag's worth of coca leaves that Silman had bought for the equivalent of about two dollars.

Standing on the mountaintop, Silman told me that the trail we were going to take down the following morning was often used by coca peddlers walking up. The *cocaleros* carried the leaves from the valleys where they are grown to high Andean villages of the sort we'd just passed, and the trail had been used for this purpose since the days of the conquistadors.

Silman, who teaches at Wake Forest University, calls himself a forest ecologist, though he also answers to the title tropical ecologist, community ecologist, or conservation biologist. He began his career thinking about how forest communities are put together, and whether they tend to remain stable over time. This led him to look at the ways the climate in the tropics had changed in the past, which led him, naturally enough, to look into how it is projected to change in the future. What he learned inspired him to establish the series of tree plots that we are about to visit. Each of Silman's plots—there are seventeen in all—sits at a different elevation and hence has a different average annual temperature. In the mega-diverse world of Manú, this means that each plot represents a slice of a fundamentally different forest community.

Silman's plots are arranged along a ridge. Plot 1, at the top of the ridge, has the highest elevation and hence the lowest annual temperature.

In the popular imagination, global warming is mostly seen as a threat to cold-loving species, and there are good reasons for this. As the world warms, the poles will be transformed. In the Arctic, perennial sea ice covers just half the area it did thirty years ago, and thirty years from now, it may well be gone entirely. Obviously, any animal that depends on the ice—ringed seals, say, or polar bears—is going to be hard-pressed as it melts away.

But global warming is going to have just as great an impact—indeed, according to Silman, an even greater impact—in the tropics. The reasons for this are somewhat more complicated, but they start with the fact that the tropics are where most species actually live.

* * *

CONSIDER for a moment the following (purely hypothetical) journey. You are standing on the North Pole one fine spring day. (There is, for the moment, still plenty of ice at the pole, so there's no danger of falling through.) You start to walk, or better yet ski. Because there is only one direction to move in, you have to go south, but you have 360 meridians to choose from. Perhaps, like me, you live in the Berkshires and are headed to the Andes, so you decide that you will follow the seventy-third meridian west. You ski and ski, and finally, about five hundred miles from the pole, you reach Ellesmere Island. All this time, of course, you will not have seen a tree or a land plant of any kind, since you are traveling across the Arctic Ocean. On Ellesmere, you will still not see any trees, at least not any that are recognizable as such. The only woody plant that grows on the island is the Arctic willow, which reaches no higher than your ankle. (The writer Barry Lopez has noted that if you spend much time wandering around the Arctic, you eventually realize "that you are standing on *top* of a forest.")

As you continue south, you cross the Nares Strait—getting around is now becoming more complicated, but we'll leave that aside—then traverse the westernmost tip of Greenland, cross Baffin Bay, and reach Baffin Island. On Baffin, there is also nothing that would really qualify as a tree, though several species of willow can be found, growing in knots close to the ground. Finally—and you are now roughly two thousand miles into your journey—you reach the Ungava Peninsula, in northern Quebec. Still you are north of the treeline, but if you keep walking for another 250 miles or so, you will reach the edge of the boreal forest. Canada's boreal forest is huge; it stretches across almost a billion acres and represents roughly a quarter of all the intact forest that remains on earth. But diversity in the boreal forest is low. Across Canada's billion acres of it, you will find only about twenty species of tree, including black spruce, white birch, and balsam fir.

Once you enter the United States, tree diversity will begin, slowly, to tick up. In Vermont, you'll hit the Eastern Deciduous Forest, which once covered almost half the country, but today remains only in patches, most of them second-growth. Vermont has something like fifty species of native trees, Massachusetts around fifty-five. North Carolina (which lies slightly to the west of your path) has more than two hundred species. Although the seventy-third meridian misses Central America altogether, it's worth noting that tiny Belize, which is about the size of New Jersey, has some seven hundred native tree species.

The seventy-third meridian crosses the equator in Colombia, then slices through bits of Venezuela, Peru, and Brazil before entering Peru again. At around thirteen degrees south latitude, it passes to the west of Silman's tree plots. In his plots, which collectively have an area roughly the size of Manhattan's Fort Tryon Park, the diversity is staggering. One thousand and thirty-five tree species have been counted there, roughly fifty times as many as in all of Canada's boreal forest.

And what holds for the trees also holds for birds and butterflies and frogs and fungi and just about any other group you can think of (though not, interestingly enough, for aphids). As a general rule, the variety of life is most impoverished at the poles and richest at low latitudes. This pattern is referred to in the scientific literature as the "latitudinal diversity gradient," or LDG, and it was noted already by the German naturalist Alexander von Humboldt, who was amazed by the biological splendors of the tropics, which offer "a spectacle as varied as the azure vault of the heavens."

"The verdant carpet which a luxuriant Flora spreads over the surface of the earth is not woven equally in all parts," Humboldt wrote after returning from South America in 1804. "Organic development and abundance of vitality gradually increase from the poles towards the equator." More than two centuries later, why this should be the case is still not known, though more than thirty theories have been advanced to explain the phenomenon.

One theory holds that more species live in the tropics because the evolutionary clock there ticks faster. Just as farmers can produce more harvests per year at lower latitudes, organisms can produce more generations. The greater the number of generations, the higher the chances of genetic mutations. The higher the chances of mutations, the greater the likelihood that new species will emerge. (A slightly different but related theory has it that higher temperatures in and of themselves lead to higher mutation rates.)

A second theory posits that the tropics hold more species because tropical species are finicky. According to this line of reasoning, what's important about the tropics is that temperatures there are relatively stable. Thus tropical organisms tend to possess relatively narrow thermal tolerances, and even slight climatic differences, caused, say, by hills or valleys, can constitute insuperable barriers. (A famous paper on this subject is titled "Why Mountain Passes Are Higher in the Tropics.") Populations are thus more easily isolated, and speciation ensues.

Yet another theory centers on history. According to this account, the most salient fact about the tropics is that they are old. A version of the Amazon rainforest has existed for many millions of years, since before there even was an Amazon. Thus, in the tropics, there's been lots of time for diversity to, as it were, accumulate. By contrast, as recently as twenty thousand years ago, nearly all of Canada was covered by ice a mile thick. So was much of New England, meaning that every species of tree now found in Nova Scotia or Ontario or Vermont or New Hampshire is a migrant that's arrived (or returned) just in the last several thousand years. The diversity as a function of time theory was first advanced by Darwin's rival, or, if you prefer, codiscoverer, Alfred Russel Wallace, who observed that in the tropics "evolution has had a fair chance," while in glaciated regions "it has had countless difficulties thrown in its way."

* * *

THE following morning, we all crawled out of our sleeping bags early to see the sunrise. Overnight, clouds had rolled in from the Amazon basin, and we watched them from above as they turned first pink and then flaming orange. In the chilly dawn, we packed up our gear and headed down the trail. "Pick out a leaf with an interesting shape," Silman instructed me once we'd descended into the cloud forest. "You'll see it for a few hundred meters, and then it will be gone. That's it. That's the tree's entire range."

Silman was carrying a two-foot-long machete, which he used to hack away at the undergrowth. Occasionally, he waved it in the air to point out something interesting: a spray of tiny white orchids with flowers no bigger than a grain of rice; a plant in the blueberry family with vivid red berries; a parasitic shrub with bright orange flowers. One of Silman's graduate students, William Farfan Rios, handed me a leaf the size of a dinner plate.

"This is a new species," he said. Along the trail, Silman and his students have found thirty species of trees new to science. (Just this grove of discoveries represents half again as many species as in Canada's boreal forest.) And there are another three hundred species that they suspect may be new, but that have yet to be formally classified. What's more, they've discovered an entirely new genus.

"That's not like finding another *kind* of oak or another *kind* of hickory," Silman observed. "It's like finding 'oak' or 'hickory.'" Leaves from trees in the genus had been sent to a specialist at the University of California-Davis, but, unfortunately, he had died before figuring out where on the taxonomic tree to stick the new branch.

Although it was winter in the Andes and the height of the dry season, the trail was muddy and slick. It had worn a deep channel into the mountainside, so that as we walked along, the ground was at eye level. At various points, trees had grown across the top and the channel became a tunnel. The first tunnel we hit was dark and dank and dripping with fine rootlets. Later tunnels were longer and darker and even in the middle of the day required a headlamp to navigate. Often I felt as if I'd entered into a very grim fairy tale.

We passed Plot 1, elevation 11,320 feet, but did not stop there. Plot 2, elevation 10,500 feet, had been recently scoured by a landslide; this pleased Silman because he was interested to see what sorts of trees would recolonize it.

The farther we descended, the denser the forest became. The trees were not just trees; they were more like botanical gardens, covered with ferns and orchids and bromeliads and strung with lianas. In some spots, the vegetation was so thick that soil mats had formed above the ground, and these had sprouted plants of their own—forests in the air. With nearly every available patch of light and bit of space occupied, the competition for resources was evidently fierce, and it almost seemed possible to watch natural

The view from Plot 4.

selection in action, "daily and hourly" scrutinizing "every variation, even the slightest." (Another theory of why the tropics are so diverse is that greater competition has pushed species to become more specialized, and more specialists can coexist in the same amount of space.) I could hear birds calling, but only rarely could I spot them; it was difficult to see the animals for the trees.

Somewhere around Plot 3, elevation 9,680 feet, Silman pulled out the shopping bag full of coca leaves. He and his students were carrying what seemed to me to be a ridiculous amount of heavy stuff: a bag of apples, a bag of oranges, a seven-hundred-page bird book, a nine-hundred-page plant book, an iPad, bottles of benzene, a can of spray paint, a wheel of cheese, a bottle of rum. Coca, Silman told me, made a heavy pack feel lighter. It also staved off hunger, alleviated aches and pains, and helped counter altitude sickness. I had been given little to carry besides my own gear; still, anything that would lighten my pack seemed worth trying. I took a handful of leaves and a pinch of baking soda. (Baking soda, or some other alkaline substance, is necessary for coca to have its pharmaceutical effect.) The leaves were leathery and tasted like old books. Soon my lips grew numb, and my aches and pains began to fade. An hour or two later, I was back for more. (Many times since have I wished for that shopping bag.)

In the early afternoon, we reached a small, soggy clearing where, I was informed, we were going to spend the night. This was the edge of Plot 4, elevation 8,860 feet. Silman and his students had often camped there before, sometimes for weeks at a stretch. The clearing was strewn with bromeliads that had been pulled down and gnawed upon. Silman identified these as the leavings of a spectacled bear. The spectacled bear, also known as the Andean bear, is South America's last surviving bear. It is black or dark brown with beige around its eyes, and it lives mainly off plants. I hadn't realized that there were bears in the Andes, and I couldn't help thinking of Paddington, arriving in London from "deepest, darkest Peru."

* * *

EACH of Silman's seventeen tree plots is two and a half acres, and the plots are arranged along a ridge a bit like buttons on a cloak. They run from the top of the ridge all the way down to the Amazon basin, which is pretty much at sea level. In the plots, someone—Silman or one of his graduate students—has tagged every single tree over four inches in diameter. Those trees have been measured, identified by species, and given a number. Plot 4 has 777 trees over four inches, and these belong to sixty different species. Silman and his students were preparing to recensus the plots, a project that was expected to take several months. All the trees that had already been tagged would have to be remeasured, and any tree that had shown up or died since the last count would have to be added or subtracted. There were long, Talmudic discussions, conducted partly in English and partly in Spanish, about how, exactly, the recensus

In the plots, each tree over four inches in diameter has been tagged.

should be conducted. One of the few that I could follow centered on asymmetry. A tree trunk is not perfectly circular, so depending on how you orient the calipers when you're measuring, you'll get a different diameter. Eventually, it was decided that the calipers should be oriented with their fixed jaw on a dot spray-painted on every tree in red.

Owing to the differences in elevation, each of Silman's plots has a different average annual temperature. For example, in Plot 4 the average is fifty-three degrees. In Plot 3, which is about eight hundred feet higher, it's fifty-one degrees, and in Plot 5, which is about eight hundred feet lower, it's fifty-six degrees. Because tropical species tend to have narrow thermal ranges, these temperature differences translate into a high rate of turnover; trees that are abundant in one plot may be missing entirely from the next one down or up.

"Some of the dominants have the narrowest altitudinal range," Silman told me. "This suggests that what makes them such good competitors in this range makes them not so good outside of it." In Plot 4, for example, ninety percent of the tree species are different from those species found in Plot 1, which is only about twenty-five hundred feet higher.

Silman first laid out the plots in 2003. His idea was to keep coming back, year after year, decade after decade, to see what happened. How would the trees respond to climate change? One possibility—what might be called the Birnam Wood scenario—was that the trees in each zone would start moving upslope. Of course, trees can't actually move, but they can do the next best thing, which is to disperse seeds that grow into new trees. Under this scenario, species now found in Plot 4 would, as the climate warmed, start appearing higher upslope, in Plot 3, while Plot 3's would appear in Plot 2, and so on. Silman and his students completed the first recensus in 2007. Silman thought of the effort as part of his long-term project and couldn't imagine that much of interest would be found after just four years. But one of his post-docs, Kenneth Feeley, insisted on sifting through all the data, anyway. Feeley's work revealed that the forest was already, measurably, in motion.

There are various ways to calculate migration rates: for instance, by the number of trees or, alternatively, by their mass. Feeley grouped the trees by genus. Very roughly speaking, he found that global warming was driving the average genus up the mountain at a rate of eight feet per year. But he also found the average masked a surprising range of response. Like cliques of kids at recess, different trees were behaving in wildly different ways.

Take, for example, trees in the genus *Schefflera*. *Schefflera*, which is part of the ginseng family, has palmately compound leaves; these are arrayed around a central point the way your fingers are arranged around your palm. (One member of the group, *Schefflera arboricola*, from Taiwan, commonly known as the dwarf umbrella tree, is often grown as a houseplant.) Trees in *Schefflera*, Feeley found, were practically hyperactive; they were racing up the ridge at the astonishing rate of nearly a hundred feet a year.

On the opposite extreme were trees in the genus *Ilex*. These have alternate leaves that are usually glossy, with spiky or serrated edges. (The genus includes *Ilex aquifolium*, which is native to Europe and known to Americans as Christmas holly.) The trees in *Ilex* were like kids who spend recess sprawled out on a bench. While *Schefflera* was sprinting upslope, *Ilex* was just sitting there, more or less inert.

ANY species (or group of species) that can't cope with some variation in temperatures is not a species (or group) whose fate we need be concerned about right now, because it no longer exists. Everywhere on the surface of the earth temperatures fluctuate. They fluctuate from day to night and from season to season. Even in the tropics, where the difference between winter and summer is minimal, temperatures can vary significantly between the rainy and the dry seasons. Organisms have developed all sorts of ways of dealing with these variations. They hibernate or estivate or migrate. They dissipate heat through panting or conserve it by growing thicker coats of fur. Honeybees warm themselves by contracting the muscles in their thorax. Wood storks cool off by defecating on their own legs. (In very hot weather, wood storks may excrete on their legs as often as once a minute.)

Over the lifetime of a species, on the order of a million years, longer-term temperature changes—changes in climate—come into play. For the last forty million years or so, the earth has been in a general cooling phase. It's not entirely clear why this is so, but one theory has it that the uplift of the Himalayas exposed vast expanses of rock to chemical weathering, and this in turn led to a drawdown of carbon dioxide from the atmosphere. At the start of this long cooling phase, in the late Eocene, the world was so warm there was almost no ice on the planet. By around thirty-five million years ago, global temperatures had declined enough that glaciers began to form on Antarctica. By three million years ago, temperatures had dropped to the point that the Arctic, too, froze over, and a permanent ice cap formed. Then, about two and a half million years ago, at the start of the Pleistocene epoch, the world entered a period of recurring glaciations. Huge ice sheets advanced across the Northern Hemisphere, only to melt away again some hundred thousand years later.

Even after the idea of ice ages was generally accepted—it was first proposed in the eighteen-thirties by Louis Agassiz, a protégé of Cuvier—no one could explain how such an astonishing process could take place. In 1898, Wallace observed that "some of the most acute and powerful intellects of our day have exerted their ingenuity" on the problem, but so far "altogether in vain." It would take another three-quarters of a century for the question to be resolved. It is now generally believed that ice ages are initiated by small changes in the earth's orbit, caused by, among other things, the gravitational tug of Jupiter and Saturn. These changes alter the distribution of sunlight across different latitudes at different times of year. When the amount of light hitting the far northern latitudes in summer approaches a minimum, snow begins to build up there. This initiates a feedback cycle that causes atmospheric carbon dioxide levels to drop. Temperatures fall, which leads more ice to build up, and so on. After a while, the orbital cycle enters a new phase, and the feedback loop begins to run in reverse. The ice starts to melt, global CO_2 levels rise, and the ice melts back farther.

During the Pleistocene, this freeze-thaw pattern was repeated some twenty times, with world-altering effects. So great was the amount of water tied up in ice during each glacial episode that sea levels dropped by some three hundred feet, and the sheer weight of the sheets was enough to depress the crust of the earth, pushing it down into the mantle. (In places like northern Britain and Sweden, the process of rebound from the last glaciation is still going on.)

How did the plants and animals of the Pleistocene cope with these temperature swings? According to Darwin, they did so by moving. In *On the Origin of Species*, he describes vast, continental-scale migrations.

> As the cold came on, and as each more southern zone became fitted for arctic beings and ill-fitted for their former more temperate inhabitants, the latter would be supplanted and arctic productions would take their places. . . . As the warmth returned, the arctic forms would retreat northward, closely followed up in their retreat by the productions of the more temperate regions.

Darwin's account has since been confirmed by all sorts of physical traces. Researchers studying ancient beetle casings, for example, have found that during the ice ages, even tiny insects migrated thousands of miles to track the climate. (To name just one of these, *Tachinus caelatus* is a small, dullish brown beetle that today lives in the mountains west of Ulan Bator, in Mongolia. During the last glacial period, it was common in England.)

In its magnitude, the temperature change projected for the

coming century is roughly the same as the temperature swings of the ice ages. (If current emissions trends continue, the Andes are expected to warm by as much as nine degrees.) But if the magnitude of the change is similar, the rate is not, and, once again, rate is key. Warming today is taking place at least ten times faster than it did at the end of the last glaciation, and at the end of all those glaciations that preceded it. To keep up, organisms will have to migrate, or otherwise adapt, at least ten times more quickly. In Silman's plots, only the most fleet-footed (or rooted) trees, like the hyperactive genus *Schefflera*, are keeping pace with rising temperatures. How many species overall will be capable of moving fast enough remains an open question, though, as Silman pointed out to me, in the coming decades we are probably going to learn the answer, whether we want to or not.

MANÚ National Park, where Silman's plots are laid out, sits in the southeastern corner of Peru, near the country's borders with Bolivia and Brazil, and it stretches over nearly six thousand square miles. According to the United Nations Environment Programme, Manú is "possibly the most biologically diverse protected area in the world." Many species can be found only in the park and its immediate environs; these include the tree fern *Cyathea multisegmenta*, a bird known as the white-cheeked tody flycatcher, a rodent called Barbara Brown's brush-tailed rat, and a small, black toad known only by its Latin name, *Rhinella manu*.

The first night on the trail, one of Silman's students, Rudi Cruz, insisted that we all go out looking for *Rhinella manu*. He had seen several of the toads during a previous visit to the spot, and he felt sure we could find them again if we tried. I'd recently read a paper on the spread of the chytrid fungus to Peru—according to the authors, it had already arrived in Manú—but I decided not to mention this. Perhaps *Rhinella manu* was still out there, in which case I certainly wanted to see it.

We strapped on headlamps and set out down the trail, like a line of coal miners filing down a shaft. The forest at night had become an impenetrable tangle of black. Cruz led the way, shining his lamp along the tree trunks and peering into the bromeliads. The rest of us followed suit. This went on for maybe an hour and turned up only a few brownish frogs from the genus *Pristimantis*. After a while, people started getting bored and drifting back to camp. Cruz refused to give up. Perhaps thinking that the problem was the rest of us, he headed up the trail in the opposite direction. "Did you find anything?" someone would periodically call out to him through the darkness.

"*Nada*," came the repeated response.

The next day, after more arcane discussions about tree measurements, we packed up to continue down the ridge. On a trip to fetch water, Silman had found a spray of white berries interspersed with what looked like bright purple streamers. He'd identified the arrangement as the inflorescence of a tree in the Brassicaceae, or mustard, family, but he had never seen anything quite like it before, which made him think, he told me, that it might represent yet another new species. It was pressed in newspaper for transport down the mountain. The idea that I might have been present at the discovery of a species, even though I'd had absolutely nothing to do with it, filled me with an odd sort of pride.

BACK on the trail, Silman hacked away with his machete, pausing every now and then to point out a new botanical oddity, like a shrub that steals water from its neighbors by sticking out needlelike roots. Silman talks about plants the way other people speak about movie stars. One tree he described to me as "charismatic." Others were "hilarious," "crazy," "neat," "clever," and "amazing."

Sometime in the mid-afternoon, we emerged onto a rise with a view across a valley to the next ridge. On the ridge, the trees were shaking. This was a sign of woolly monkeys making their way through the forest. Everyone stopped to try to get a glimpse of them. As they sailed from branch to branch, the monkeys made a chirruping noise, a bit like the whine of crickets. Silman pulled out the shopping bag and passed it around.

A little while later, we reached Plot 6, elevation 7,308 feet, where the tree from the new genus had been found. Silman waved his machete at it. The tree looked pretty ordinary, but I tried to see it through his eyes. It was taller than most of its neighbors—perhaps it could be described as "stately" or "statuesque"—with smooth, ruddy bark and simple, alternate leaves. It belonged to the Euphorbiaceae, or spurge, family, whose members include poinsettia. Silman was eager to learn as much as possible about the tree, so that when a new taxonomist could be found to replace the one who had died, he'd be able to send him all the necessary material. He and Farfan went to see what they could come up with. They returned with some seed capsules, which were as thick and tough as hazelnut shells, but delicately shaped, like flowering lilies. The capsules were dark brown on the outside and inside the color of sand.

That evening, the sun set before we reached Plot 8, where we were going to camp. We hiked on through the dark, then set up our tents and made dinner, also in the dark. I crawled into my sleeping bag around 9 PM, but a few hours later, I was woken by a light. I assumed someone had gotten up to pee, and rolled over. In the morning, Silman told me that he was surprised I'd been able to sleep through all the commotion. Six groups of *cocaleros* had tromped through the campsite overnight. (In Peru, though the sale of coca is legal, all purchases are supposed to go through a government agency known as ENACO, a restriction growers do their best to avoid.) Every single group had tripped over his tent. Eventually he'd gotten so annoyed, he'd yelled at the *cocaleros*, which, he had to admit, probably hadn't been the wisest idea.

* * *

In ecology, rules are hard to come by. One of the few that's universally accepted is the "species-area relationship," or SAR, which has been called the closest thing the discipline has to a periodic table. In its broadest formulation, the species-area relationship seems so simple as to be almost self-evident. The larger the area you sample, the greater the number of species you will encounter. This pattern was noted all the way back in the seventeen-seventies by Johann Reinhold Forster, a naturalist who sailed with Captain Cook on his second voyage, the one after his unfortunate collision with the Great Barrier Reef. In the nineteen-twenties, it was codified mathematically by a Swedish botanist, Olof Arrhenius. (As it happens, Olof was the son of the chemist Svante Arrhenius, who, in the eighteen-nineties, showed that burning fossil fuels would lead to a warmer planet.) And it was further refined and elaborated by E. O. Wilson and his colleague Robert MacArthur in the nineteen-sixties.

The correlation between the number of species and the size of the area is not linear. Rather, it's a curve that slopes in a predictable way. Usually, the relationship is expressed by the formula $S = cA^z$,

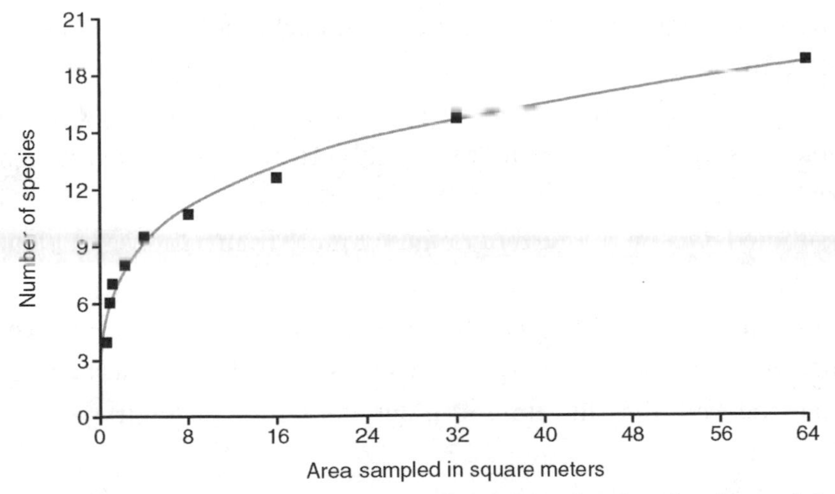

A typical example of the species-area relationship, showing the shape of the curve.

where S is the number of species, A is the size of the area, and c and z are constants that vary according to the region and taxonomic group under consideration (and hence are not really constants in the usual sense of the term). The relationship counts as a rule because the ratio holds no matter what the terrain. You could be studying a chain of islands or a rainforest or a nearby state park, and you'd find that the number of species varies according to the same insistent equation: $S = cA^z$.*

For the purposes of thinking about extinction, the species-area relationship is key. One (admittedly simplified) way of conceiving of what humans are doing to the world is that we are everywhere changing the value of A. Consider, for example, a grassland that once covered a thousand square miles. Let's say the grassland was home to a hundred species of birds (or beetles or snakes). If half of the grassland were eliminated—converted into farmland or shopping malls—it should be possible to calculate, using the species-area relationship, the proportion of bird species (or beetles or snakes) that would be lost. Very roughly speaking, the answer is ten percent. (Here again, it's important to remember that the relationship is not linear.) Since it takes a long time for the system to reach a new equilibrium, you wouldn't expect the species to disappear right away, but you would expect them to be headed in that direction.

In 2004, a group of scientists decided to use the species-area relationship to generate a "first-pass" estimate of the extinction risk posed by global warming. First, the members of the team gathered data on the current ranges of more than a thousand plant and animal species. Then they correlated these ranges with present-day climate conditions. Finally, they imagined two extreme scenarios. In one, all of the species were assumed to be inert, much like the *Ilex* trees in Silman's plots. As temperatures rose, they

*It's important to note that z is always less than 1—usually somewhere between .20 and .35.

stayed put, and so, in most cases, the amount of climatically suitable area available to them shrank, in many instances down to zero. The projections based on this "no dispersal" scenario were bleak. If warming were held to a minimum, the team estimated that between 22 and 31 percent of the species would be "committed to extinction" by 2050. If warming were to reach what was at that point considered a likely maximum—a figure that now looks too low—by the middle of this century, between 38 and 52 percent of the species would be fated to disappear.

"Here's another way to express the same thing," Anthony Barnosky, a paleontologist at the University of California-Berkeley, wrote of the study results. "Look around you. Kill half of what you see. Or if you're feeling generous, just kill about a quarter of what you see. That's what we could be talking about."

In the second, more optimistic scenario, species were imagined to be highly mobile. Under this scenario, as temperatures climbed, creatures were able to colonize any new areas that met the climate conditions they were adapted to. Still, many species ended up with nowhere to go. As the earth warmed, the conditions they were accustomed to simply disappeared. (The "disappearing climates" turned out to be largely in the tropics.) Other species saw their habitat shrink because to track the climate they had to move upslope, and the area at the top of a mountain is smaller than at the base.

Using the "universal dispersal" scenario, the team, led by Chris Thomas, a biologist at the University of York, found that, with the minimum warming projected, 9 to 13 percent of all species would be "committed to extinction" by 2050. With maximum warming, the numbers would be 21 to 32 percent. Taking the average of the two scenarios, and looking at a mid-range warming projection, the group concluded that 24 percent of all species would be headed toward extinction.

The study ran as the cover article in *Nature*. In the popular press, the welter of numbers the researchers came up with was condensed down to just one. "Climate Change Could Drive a Million of the World's Species to Extinction," the BBC declared. "By 2050 Warming to Doom a Million Species" is how the headline in *National Geographic* put it.

The study has since been challenged on a number of grounds. It ignores interactions between organisms. It doesn't account for the possibility that plants and animals can tolerate a broader range of climates than their current range suggests. It looks only as far as 2050 when, under any remotely plausible scenario, warming will continue far beyond that. It applies the species-area relationship to a new, and therefore untested, set of conditions.

More recent studies have come down on both sides of the *Nature* paper. Some have concluded that the paper overestimated the number of extinctions likely to be caused by climate change, others that it understated it. For his part, Thomas has acknowledged that many of the objections to the 2004 paper may be valid. But he has pointed out that every estimate that's been proposed since then has been the same order of magnitude. Thus, he's observed, "around 10 or more percent of species, and not 1 percent, or .01 percent," are likely to be done in by climate change.

In a recent article, Thomas suggested that it would be useful to place these numbers "in a geological context." Climate change alone "is unlikely to generate a mass extinction as large as one of the Big Five," he wrote. However, there's a "high likelihood that climate change on its own could generate a level of extinction on par with, or exceeding, the slightly 'lesser' extinction events" of the past.

"The potential impacts," he concluded, "support the notion that we have recently entered the Anthropocene."

"THE Brits like to mark everything in plastic," Silman told me. "We think it's kind of gauche." It was our third day on the trail, and we were standing in Plot 8, where we'd come across a strip of blue tape outlining the plot's border. Silman suspected that it was the handiwork of colleagues of his from Oxford. Silman spends a lot of time in Peru—sometimes months at a stretch—but much of the year he's not there, and all sorts of things can happen that he doesn't know about (and, usually, doesn't care for). For instance, on our trip Silman found several wire baskets that had been suspended in the tree plots to catch seeds. Clearly, they'd been set up for research purposes, but no one had told him about them or asked his permission, and so they represented a sort of scientific piracy. I imagined rogue researchers creeping through the forest like *cocaleros*.

In Plot 8, Silman introduced me to another "really interesting" tree, *Alzatea verticillata*. *Alzatea verticillata* is unusual in that it is the only species in its genus, and even more unusual in that it's the only species in its family. It has papery, bright green, oblong leaves and small white flowers which, according to Silman, smell like burnt sugar when in bloom. *Alzatea verticillata* can grow to be very tall, and at this particular elevation—around fifty-nine hundred feet—it is the dominant canopy tree in the forest. It is one of those species that seem to be just sitting there motionless.

Silman's plots represent another response to Thomas—one that's practical rather than theoretical. Trees are obviously a lot less mobile than, say, trogons—tropical birds common in Manú—or even ticks. But in a cloud forest, trees structure the ecosystem, much as corals structure a reef. Certain types of insects depend on certain types of trees, and certain sorts of birds depend on those insects, and so on up the food chain. The reverse is also true: animals are critical to the survival of the forest. They are the pollinators and seed dispersers, and the birds prevent the insects from taking over. At the very least, Silman's work suggests, global warming will restructure ecological communities. Different groups of trees will respond differently to warming, and so contemporary associations will break down. New ones will form. In this planet-wide restructuring, some species will thrive. Many plants may in fact benefit from high carbon dioxide levels, since it will be easier for them to obtain the CO_2 they need for photosynthesis. Others will fall behind and eventually drop out.

Silman sees himself as an upbeat person. This is—or at least was—reflected in his research. "My lab has kind of been the sunshine lab," he told me. He has argued publicly that with better policing and well-placed reserves, many threats to biodiversity—illegal logging, mining, ranching—could be minimized.

"Even in tropical areas, we know how to stop this stuff," he said. "We're getting better governance."

But in a rapidly warming world, the whole idea of a well-placed reserve becomes, if not exactly moot, then certainly a lot more problematic. In contrast to, say, a logging crew, climate change cannot be forced to respect a border. It will alter the conditions of life in Manú just as surely as it will alter them in Cuzco or Lima. And with so many species on the move, a reserve that's fixed in place is no stay against loss.

"This is a qualitatively different set of stresses that we are putting on species," Silman told me. "In other kinds of human disturbances there were always spatial refuges. Climate affects *everything*." Like ocean acidification, it is a global phenomenon, or, to borrow from Cuvier, a "revolution on the surface of the earth."

THAT afternoon, we emerged onto a dirt road. Silman had collected various plants that interested him to take back to his lab, and these were strapped to his enormous backpack, so that he resembled a cloud-forest Johnny Appleseed. The sun was out, but it had recently rained, and clusters of black and red and blue butterflies hovered over the puddles. Occasionally, a truck rumbled by, loaded down with logs. The butterflies couldn't scatter fast enough, so the road was littered with severed wings.

We walked until we reached a clutch of tourist lodges. The area we'd entered, Silman told me, was famous among birders, and just trudging along the road, we saw a rainbow assortment of species: golden tanagers the color of buttercups, blue-gray tanagers the color of cornflowers, and blue-necked tanagers, which are a flash of dazzling turquoise. We also saw a silver-beaked tanager with a bright red belly and a flock of Andean cock-of-the-rocks, known for their flamboyant scarlet feathers. Male cock-of-the-rocks have a disk-shaped crest on the top of their heads and a raspy call that makes them sound demented.

At various points in earth history, the sorts of creatures now restricted to the tropics had much broader ranges. During the mid-Cretaceous, for example, which lasted from about 120 to 90 million years ago, breadfruit trees flourished as far north as the Gulf of Alaska. In the early Eocene, about 50 million years ago, palms grew in the Antarctic, and crocodiles paddled in the shallow seas around England. There's no reason to suppose, in the abstract, that a warmer world would be any less diverse than a colder one; on the contrary, several possible explanations for the "latitudinal diversity gradient" suggest that, over the long term, a warmer world would be *more* varied. In the short term, though, which is to say, on any timescale that's relevant to humans, things look very different.

Virtually every species that's around today can be said to be cold-adapted. Golden tanagers and cock-of-the-rocks, not to mention bluejays and cardinals and barn swallows, all made it through the last ice age. Either they or their very close relatives also made it through the ice age before that, and the one before that, and so on going back two and a half million years. For most of the Pleistocene temperatures were significantly lower than they are now—such is the rhythm of the orbital cycle that glacial periods tend to last much longer than interglacials—and so an evolutionary premium was placed on being able to deal with wintry conditions. Meanwhile, for two and a half million years, there's been no advantage in being able to deal with extra heat, since temperatures never got much warmer than they are right now. In the ups and downs of the Pleistocene, we are at the crest of an up.

To find carbon dioxide levels (and therefore, ultimately, global temperatures) higher than today's requires going back a long way, perhaps as far as the mid-Miocene, fifteen million years ago. It's quite possible that by the end of this century, CO_2 levels could reach a level not seen since the Antarctic palms of the Eocene, some fifty million years ago. Whether species still possess the features that allowed their ancestors to thrive in that ancient, warmer world is, at this point, impossible to say.

"For plants to tolerate warmer temperatures there's all sorts of things that they could do," Silman told me. "They could manufacture special proteins. They could change their metabolism, things like that. But thermal tolerance can be costly. And we haven't seen temperatures like those that are predicted in millions of years. So the question is: have plants and animals retained over this huge amount of time—whole radiations of mammals have come and gone in this period—have they retained these potentially costly characteristics? If they have, then we may get a pleasant surprise." But what if they haven't? What if they've lost these costly characteristics because for so many millions of years they provided no advantage?

"If evolution works the way it usually does," Silman said, "then the extinction scenario—we don't call it extinction, we talk about it as 'biotic attrition,' a nice euphemism—well, it starts to look apocalyptic."

CHAPTER IX

ISLANDS ON DRY LAND

Eciton burchellii

BR-174 RUNS FROM THE CITY OF MANAUS, IN THE BRAZILIAN state of Amazonas, more or less due north to the Venezuelan border. The road used to be lined with the wreckage of cars that had skidded off to one side or the other, but since it was paved, about twenty years ago, it has become easier to navigate and now, instead of burned-out hulks, there's an occasional café catering to travelers. After an hour or so the cafés give out, and after another hour, there's a turnoff to a single-lane road, ZF-3, that heads due east. ZF-3 remains unpaved, and, owing to the color of the dirt in Amazonas, it appears as a bright orange gash tearing through the countryside. Follow ZF-3 for another three-quarters of an hour and you reach a wooden gate closed with a length of chain. Beyond the gate, some cows are standing around looking sleepy, and beyond the cows is what's known as Reserve 1202.

Reserve 1202 might be thought of as an island at the center of the Amazon. I arrived there on a hot, cloudless day in the middle of the rainy season. Fifty feet into the reserve, the foliage was so dense that even with the sun directly overhead, the light was still murky, as in a cathedral. From a nearby tree came a high-pitched squeal that made me think of a police whistle. This, I was told, was the call of a small, unassuming bird known as a screaming piha. The piha screamed again, then fell silent.

Unlike a naturally occurring island, Reserve 1202 is an almost perfect square. It is twenty-five acres of untouched rainforest surrounded by a "sea" of scrub. In aerial photos it shows up as a green raft bobbing on waves of brown.

Reserve 1202 is part of a whole archipelago of Amazonian islands, all with equally clinical-sounding names: Reserve 1112, Reserve 1301, Reserve 2107. Some of the reserves are even smaller than twenty-five acres; a few are quite a bit bigger. Collectively, they represent one of the world's largest and longest-running experiments, the Biological Dynamics of Forest Fragments Project or, for short, the BDFFP. Pretty much every square foot of the BDFFP has been studied by someone: a botanist tagging trees, an ornithologist banding birds, an entomologist counting fruit flies. When I visited Reserve 1202, I ran into a graduate student from Portugal who was surveying bats. At noon he had just recently woken up and was eating pasta in a shed that served as a research station-cum-kitchen. While we were talking, a very skinny cowboy rode up on an only slightly less skinny horse. He had a rifle slung over one shoulder. I wasn't sure whether he'd come because he'd heard the truck I'd arrived on and wanted to protect the student from possible intruders, or because he sensed that there was pasta.

The BDFFP is the result of an unlikely collaboration between cattlemen and conservationists. In the nineteen-seventies, the Brazilian government set out to encourage ranchers to settle north of Manaus, an area that was then largely uninhabited. The program amounted to subsidized deforestation: any ranchers who agreed to move to the rainforest, cut down the trees, and start raising cows would get a stipend from the government. At the same time, under

Brazilian law, landholders in the Amazon had to leave intact at least half the forest on their property. The tension between these two directives gave an American biologist named Tom Lovejoy an idea. What if the ranchers could be convinced to let scientists decide which trees to cut down and which ones to leave standing? "The idea was really just one sentence," Lovejoy told me. "I wondered if you could persuade the Brazilians to arrange the fifty percent so you could have a giant experiment." In that case, it would be possible to study in a controlled way a process that was taking place in an uncontrolled fashion all across the tropics, indeed across the entire world.

Lovejoy flew to Manaus and presented his plan to Brazilian officials. Rather to his surprise, they embraced it. The project has now been running continuously for more than thirty years. So many graduate students have been trained at the reserves that a new word was coined to describe them: "fragmentologist." For its

Forest fragments north of Manaus, as seen from the air.

part, the BDFFP has been called "the most important ecological experiment ever done."

CURRENTLY, about fifty million square miles of land on the planet are ice-free, and this is the baseline that's generally used for calculating human impacts. According to a recent study published by the Geological Society of America, people have "directly transformed" more than half of this land—roughly twenty-seven million square miles—mostly by converting it to cropland and pasture, but also by building cities and shopping malls and reservoirs, and by logging and mining and quarrying. Of the remaining twenty-three million square miles, about three-fifths is covered by forest—as the authors put it, "natural but not necessarily virgin"—and the rest is either high mountains or tundra or desert. According to another recent study, published by the Ecological Society of America, even such dramatic figures understate our impact. The authors of the second study, Erle Ellis of the University of Maryland and Navin Ramankutty of McGill, argue that thinking in terms of biomes defined by climate and vegetation—temperate grasslands, say, or boreal forests—no longer makes sense. Instead, they divide the world up into "anthromes." There is an "urban" anthrome that stretches over five hundred thousand square miles, an "irrigated cropland" anthrome (a million square miles), and a "populated forest" (four and a half million square miles). Ellis and Ramankutty count a total of eighteen "anthromes," which together extend over thirty-nine million square miles. This leaves outstanding some eleven million square miles. These areas, which are mostly empty of people and include stretches of the Amazon, much of Siberia and northern Canada, and significant expanses of the Sahara, the Gobi, and the Great Victoria deserts, they call "wildlands."

But in the Anthropocene it's not clear that even such "wildlands" really deserve to be called wild. Tundra is crisscrossed by pipelines, boreal forest by seismic lines. Ranches and plantations and hydroelectric projects slice through the rainforest. In Brazil, people speak of the "fishbone," a pattern of deforestation that begins with the construction of one major road—by this metaphor, the spine—that then leads to the creation (sometimes illegal) of lots of smaller, riblike roads. What's left is a forest of long, skinny patches. These days every wild place has, to one degree or another, been cut into and cut off. And this is what makes Lovejoy's forest fragment experiment so important. With its square, completely unnatural outline, Reserve 1202 represents, increasingly, the shape of the world.

THE cast at the BDFFP is constantly changing, so even people who have worked on the project for many years are not quite sure whom they're going to bump into there. I drove out to Reserve 1202 with Mario Cohn-Haft, an American ornithologist who first got involved with the project as an intern in the mid–nineteen-eighties. Cohn-Haft ended up marrying a Brazilian and now has a job at the National Institute of Amazonian Research in Manaus. He is tall and narrow, with wispy gray hair and mournful brown eyes. The kind of affection and enthusiasm Miles Silman brings to tropical trees, Cohn-Haft saves for birds. At one point I asked him how many Amazonian bird species he could identify by their calls, and he gave me a quizzical look, as if he didn't understand what I was getting at. When I restated the question, the answer turned out to be all of them. By the official count, there are something like thirteen hundred species of birds in the Amazon, but Cohn-Haft thinks there are actually a good many more, because people have relied too much on features like size and plumage and not paid enough attention to sound. Birds that might look more or less identical but produce different calls often turn out, he told me, to be genetically distinct. At the time of our trip, Cohn-Haft was getting ready to publish a paper identifying several new species he had discovered through rigorous listening. One of these, a nocturnal bird in the potoo family, has a sad, haunting call, which locals sometimes attribute to the *curupira*, a figure from Brazilian folklore. The *curupira* has a boyish face, copious hair, and backward-pointing feet. He preys on poachers and anyone else who takes too much from the forest.

Because dawn is the best time to hear birds, Cohn-Haft and I set out for Reserve 1202 in the dark, shortly after 4 AM. Our first stop, along the way, was a metal tower built to support a weather station. From the top of the tower, which was about 130 feet high and in an advanced state of rust, there was a panoramic view over the forest canopy. Cohn-Haft had brought along a powerful scope, which he set up on a tripod. He'd also brought along an iPod and a miniature loudspeaker that fit in his pocket. The iPod was loaded with recordings of hundreds of calls, and sometimes when he heard a bird he couldn't locate, he would play its song in the hope that it would reveal itself.

"By the end of the day you could have heard a hundred and fifty species of birds and only seen ten," he told me. Occasionally, there was a glint of color against the green, and in this way I managed to glimpse what Cohn-Haft identified as a yellow-tufted woodpecker, a black-tailed tityra, and a golden-winged parakeet. He trained the scope on a speck of blue that turned out to be the most beautiful bird I have ever seen: a red-legged honeycreeper, with a sapphire breast, scarlet legs, and a cap of brilliant aquamarine.

As the sun rose higher and the calls grew less frequent, we set out again. By the time the day had turned furnace-like and we were both dripping with sweat, we got to the chained gate that marks the entrance to Reserve 1202. Cohn-Haft chose one of the paths that have been cut into the reserve for access, and we tromped to what he thought was roughly the square's center. He stopped to listen. There wasn't much to hear.

"Right now I'm hearing only two bird species," he told me. "One of them sounds like it's saying, 'Whoops, looks like rain,' and that's a plumbeous pigeon. It's a classic primary forest species. The other is doing this 'choodle, choodle, peep' kind of thing." He made a sound like a flutist doing warm-up exercises. "And that's a rufous-browed peppershrike. And that's a typical second-growth or edge-of-pasture species that we wouldn't hear in primary forest."

Cohn-Haft explained that when he had first worked at Reserve 1202 his job was to catch and band birds and then release them, a process known by the shorthand "ring and fling." The birds were caught in nets strung across the forest from the ground to a height of six feet. Bird censuses were conducted before the forest fragments were isolated and then afterward, so that the numbers could be compared. Across the reserves—there are eleven in total—Cohn-Haft and his colleagues banded nearly twenty-five thousand birds.

"The first result that kind of surprised everyone, although it's sort of trivial in the grand scheme of things, was kind of a refugee effect," he said, as we stood in the shadows. "What happened when you cut down the surrounding forest is that the capture rate—just the number of birds you captured and the number of species sometimes, too—went up for about the first year." Apparently, birds from the deforested areas were seeking shelter in the fragments. But gradually as time went on, both the number and the variety of birds in the fragments started to drop. And then it kept on dropping.

"In other words," Cohn-Haft said, "there wasn't just suddenly this new equilibrium with fewer species. There was this steady degradation in the diversity over time." And what went for birds went for other groups as well.

ISLANDS—WE are talking about real islands now, rather than "islands" of habitat—tend to be species-poor, or, to use the term of art, depauperate. This is true of volcanic islands situated in the middle of the ocean, and it is also, more intriguingly, true of so-called land-bridge islands that are located close to shore. Researchers who have studied land-bridge islands, which are created by fluctuating sea levels, have consistently found that they are less diverse than the continents they once were part of.

Why is this so? Why should diversity drop off with isolation? For some species, the answer seems pretty straightforward: the slice of the habitat they've been marooned on is inadequate. A big cat that requires a range of forty square miles isn't likely to make it for long in an area of only twenty square miles. A tiny frog that lays its eggs in a pond and feeds on a hillside needs both a pond and a hillside to survive.

But if a lack of suitable habitat were the only issue, land-bridge islands should pretty quickly stabilize at a new, lower level of diversity. Yet they don't. They keep on bleeding species—a process that's known by the surprisingly sunny term "relaxation." On some land-bridge islands that were created by rising sea levels at the end of the Pleistocene, it's been estimated that full relaxation took thousands of years; on others, the process may still be going on.

Ecologists account for relaxation by observing that life is random. Smaller areas harbor smaller populations, and smaller populations are more vulnerable to chance. To use an extreme example, an island might be home to a single breeding pair of birds of species X. One year, the pair's nest is blown out of a tree in a hurricane. The following year, all the chicks turn out to be males, and the year after that, the nest is raided by a snake. Species X is now headed toward local extinction. If the island is home to two breeding pairs, the odds that both will suffer such a string of fatal bad luck is lower, and if it's home to twenty pairs, it's a great deal lower. But low odds in the long run can still be deadly. The process might be compared to a coin toss. It's unlikely that a coin is going to come up heads ten times in a row the first ten (or twenty or hundred) times it is flipped. However, if it's tossed often enough, even an unlikely sequence is likely to occur. The rules of probability are so robust that empirical evidence of the risks of small population size is hardly necessary; nevertheless, it's available. In the nineteen-fifties and sixties, bird-watchers kept meticulous records of every pair that bred on Bardsey Island, off Wales, from common house sparrows and oystercatchers to much rarer plovers and curlews. In the nineteen-eighties, these records were analyzed by Jared Diamond, who at that time was working as an ornithologist, specializing in the birds of New Guinea. Diamond found that the odds that any particular species had gone missing from the island could be plotted along a curve whose slope declined exponentially as the number of pairs increased. Thus, he wrote, the main predictor of local extinction was "small population size."

Small populations, of course, aren't confined to islands. A pond may have a small population of frogs, a meadow a small population of voles. And in the ordinary course of events, local extinctions occur all the time. But when such an extinction follows from a run of bad luck, the site is likely to be recolonized by members of other, more fortunate populations wandering in from somewhere else. What distinguishes islands—and explains the phenomenon of relaxation—is that recolonization is so difficult, in many cases, effectively impossible. (While a land-bridge island may support a small remnant population of, say, tigers, if that population winks out, new tigers presumably aren't going to paddle over.) The same holds true for any sort of habitat fragment. Depending on what surrounds the fragment, species may or may not be able to recolonize it once a population has been lost. Researchers at the BDFFP have found, for example, that some birds, such as white-crowned manakins, will readily cross road clearings, while others, such as scale-backed antbirds, are extremely reluctant to do so. In the absence of recolonization, local extinctions can become regional and then, eventually, global.

* * *

SOME ten miles from Reserve 1202 the dirt road peters out, and a stretch of rainforest that counts by contemporary standards as undisturbed begins. Researchers at the BDFFP have marked off sections of this forest to use as control plots, so they can compare what's happening in the fragments to what's going on in the continuous forest. Near the end of the road, there's a small camp, known as Camp 41, where they sleep and eat and try to escape from the rain. I arrived there with Cohn-Haft one afternoon just as the sky opened up. We jogged through the forest, but it really didn't matter; by the time we got to Camp 41, we were drenched.

Later, after the downpour had stopped and we'd squeezed out our socks, we headed away from the camp, deeper into the forest. The sky was still overcast, and in the gray, there was a dark and somber tint to all the greenery. I thought of the *curupira,* lurking in the trees on his backward feet.

E. O. Wilson, who visited the BDFFP twice, wrote after one of his trips, "The jungle teems, but in a manner mostly beyond the reach of the human senses." Cohn-Haft told me much the same thing, if somewhat less grandiloquently; the rainforest, he said, "looks a lot better on TV." At first it seemed to me there was nothing moving anywhere around us, but then Cohn-Haft began pointing out the signs of insect life and I began to see lots of activity going on in, to use Wilson's phrase, the "little world underneath." A stick bug hung from a dead leaf, waving its delicate legs. A spider crouched on a hoop-shaped web. A phallic tube of mud sticking up from the forest floor turned out to be the home of a cicada larva. What looked like a monstrous pregnancy bulging from a tree trunk was revealed to be a nest filled with termites. Cohn-Haft recognized a plant known as a melastome. He turned over one of its leaves and tapped on the stem, which was hollow. Tiny black ants poured out, looking as ferocious as tiny black ants can look. The ants, he explained, protect the plant from other insects in return for receiving free lodging.

Cohn-Haft grew up in western Massachusetts, as it happens not far from where I live. "Back home, I thought of myself as a general naturalist," he told me. He could name most of the trees and the insects he came across in western New England, in addition to all of the birds. But in the Amazon it was impossible to be a generalist; there was just too much to keep track of. In the BDFFP's study plots, some fourteen hundred species of trees have been identified, even more than in Silman's plots, a thousand miles to the west.

"These are megadiverse ecosystems, where every single species is very, very specialized," Cohn-Haft told me. "And in these ecosystems there's a huge premium on doing exactly what you do." He offered his own theory for why life in the tropics is so various, which is that diversity tends to be self-reinforcing. "A natural corollary to high species diversity is low population density, and that's a recipe for speciation—isolation by distance," he explained. It's also, he added, a vulnerability, since small, isolated populations are that much more susceptible to extinction.

The sun was starting to sink, and in the forest it was already twilight. As we were heading back toward Camp 41, we came upon a troop of ants following a path of their own just a few feet from ours. The reddish-brown ants were moving roughly in a straight line that led over a (to them especially) large log. They marched up the log and then down again. I followed the column as far as I could in both directions, but it seemed to go on and on and on, like a Soviet-style parade. The column, Cohn-Haft told me, consisted of army ants that belonged to the species *Eciton burchellii*.

Army ants—there are dozens of species in the tropics—differ from most other ants in that they have no fixed home. They spend their time either on the move, hunting for insects, spiders, and the occasional small lizard, or camped out in temporary "bivouacs."

An army ant from the species *Eciton burchellii*.

(*Eciton burchellii* "bivouacs" are made up of the ants themselves, arrayed around the queen in a vicious, stinging ball.) Army ants are famously voracious; a colony on the march can consume thirty thousand prey—mostly the larvae of other insects—per day. But in their very rapacity, they support a host of other species. There's a whole class of birds known as obligate ant-followers. These are almost always found around ant swarms, eating insects the ants have flushed out of the leaf litter. Other birds are opportunistic ant-followers and peck around the ants when, by chance, they encounter them. After the ant-following birds trail a variety of other creatures that are also experts at "doing exactly what they do." There are butterflies that feed on the birds' droppings and parasitic flies that deposit their young on startled crickets and cockroaches. Several species of mites hitch rides aboard the ants themselves; one species fastens itself to the ants' legs, another to its mandibles. A pair of American naturalists, Carl and Marian Rettenmeyer, who spent more than half a century studying *Eciton burchellii*, came up with a list of more than three hundred species that live in association with the ants.

Cohn-Haft didn't hear any birds and it was getting late, so we headed back to camp. We agreed that we would return to the same spot the next day to try to catch the ant-bird-butterfly procession.

In the late nineteen-seventies, an entomologist named Terry Erwin was working in Panama when someone asked him how many species of insects he thought could be found in a couple of acres of tropical forest. Up to then, Erwin had mostly been a beetle counter. He'd been spraying the tops of trees with insecticide, then collecting the carcasses that showered down from the leaves in a brittle rain. Intrigued by the larger question of how many insect species there were in the tropics as a whole, he thought about how he might extrapolate from his own experience. From a single species of tree, *Luehea seemannii,* he had collected beetles belonging to more than 950 species. Figuring that about a fifth of these beetles depended on *Luehea seemannii,* that other beetles similarly depended on other trees, that beetles represent about forty percent of all insect species, and that there are roughly fifty thousand species of tropical trees, Erwin estimated that the tropics were home to as many as thirty million species of arthropods. (In addition to insects, the group includes spiders and centipedes.) He was, he acknowledged, "shocked" by his own conclusion.

Since then, many efforts have been made to refine Erwin's estimates. Most have tended to revise the numbers downward. (Among other things, Erwin probably overstated the proportion of insects dependent on a single host plant.) Still, by all accounts, the figure remains shockingly high: recent estimates suggest there are at least two million tropical insect species and perhaps as many as seven million. By comparison, there are only about ten thousand species of birds in the entire world and only fifty-five hundred species of mammals. Thus for every species with hair and mammary glands, there are, in the tropics alone, at least three hundred with antennae and compound eyes.

The richness of its insect fauna means that any threat to the tropics translates into very high numbers of potential victims. Consider the following calculation. Tropical deforestation is notoriously difficult to measure, but let's assume that the forests are being felled at a rate of one percent annually. Using the species-area relationship, $S = cA^z$, and setting the value of z at .25, we can calculate that losing one percent of the original area implies the loss of roughly a quarter of a percent of the original species. If we assume, very conservatively, that there are two million species in the tropical rainforests, this means that something like five thousand species are being lost each year. This comes to roughly fourteen species a day, or one every hundred minutes.

This exact calculation was performed by E. O. Wilson in the late nineteen-eighties, not long after one of his trips to the BDFFP. Wilson published the results in *Scientific American*, and on the basis of them he concluded that the contemporary extinction rate was "on the order of 10,000 times greater than the naturally occurring background rate." This, he further observed, was "reducing biological diversity to its lowest level" since the end-Cretaceous extinction, an event, he noted, that while not the worst mass extinction in history, was "by far the most famous, because it ended the age of the dinosaurs, conferred hegemony on the mammals and ultimately, for better or worse, made possible the origin of our own species."

Like Erwin's, Wilson's calculations were shocking. They were also easy to grasp, or at least to repeat, and they received a great deal of attention, not just in the relatively small world of tropical biologists but also in the mainstream media. "Hardly a day passes without one being told that tropical deforestation is extinguishing roughly one species every hour, or maybe even one every minute," a pair of British ecologists lamented. Twenty-five years later, it's now generally agreed that Wilson's figures—here again like Erwin's—don't match observation, a fact that should be chastening to science writers perhaps even more than to scientists. What the reasons are for this continue to be debated.

One possibility is that extinction takes time. Wilson's calculations assume that once an area is deforested, species drop out more or less immediately. But it may take quite a while for a forest to fully "relax," and even small, remnant populations can persist for a long time, depending on the roll of the survival dice. The difference between the number of species that have been doomed by some sort of environmental change and the number that have actually vanished is often referred to as the "extinction debt." The term implies there's a lag to the process, just as there is to buying on credit.

Another possible explanation is that habitat lost to deforestation isn't really lost. Even forests that have been logged for timber or burned for pasture can and do regrow. Ironically enough, a good illustration of this comes from the area right around the BDFFP. Not long after Lovejoy convinced Brazilian officials to back the project, the country suffered a paralyzing debt crisis, and by 1990 the inflation rate was running at thirty thousand percent. The government canceled the subsidies that had been promised the ranchers, and thousands of acres were abandoned. Around some of the BDFFP's square fragments, the trees grew back so vigorously that the plots would have been swallowed up entirely had Lovejoy not arranged to have them re-isolated by cutting and burning the new growth. Though primary forest continues to decline in the tropics, secondary forest in some regions is on the rise.

Yet another possible explanation for why observations don't match predictions is that humans aren't very observant. Since the majority of species in the tropics are insects and other invertebrates, so, too, are the majority of anticipated extinctions. But as we don't know, even to the nearest million, how many tropical insect species there are, we're not likely to notice if one or two or even ten thousand of them have vanished. A recent report by the Zoological

Society of London notes that "the conservation status of less than one percent of all described invertebrates is known," and the vast majority of invertebrates probably have not yet even been described. Invertebrates may, as Wilson has put it, be "the little things that run the world," but little things are easy to overlook.

By the time Cohn-Haft and I got back to Camp 41, several other people had arrived, including Cohn-Haft's wife, Rita Mesquita, who's an ecologist, and Tom Lovejoy, who was in Manaus attending a meeting of a group called the Amazonas Sustainable Foundation. Now in his early seventies, Lovejoy is credited with having put the term "biological diversity" into general circulation and with having conceived of the idea of the "debt-for-nature swap." Over the years, he has worked for the World Wildlife Fund, the Smithsonian, the United Nations Foundation, and the World Bank, and in good part owing to his efforts something like half the Amazon rainforest is now under some form of legal protection. Lovejoy is the rare sort of person who seems equally comfortable slogging through the forest and testifying in front of Congress. He is always looking for ways to drum up support for Amazon conservation, and while we were sitting around that evening, he told me he'd once brought Tom Cruise to Camp 41. Cruise, he said, had seemed to enjoy himself, but, unfortunately, had never taken up the cause.

By now, more than five hundred scientific papers and several scientific books have been written about the BDFFP. When I asked Lovejoy to sum up what had been learned from the project, he said that one had to be cautious extrapolating from a part to the whole. For example, recent work has shown that changes in land use in the Amazon also affect atmospheric circulation. This means that, on a large enough scale, destruction of the rainforest could result not just in a disappearance of the forest but in a disappearance of the rain.

"Suppose you ended up with a landscape cut up into hundred-hectare fragments," Lovejoy said. "I think what the project has shown is that you basically would have lost more than half the fauna and flora. Of course, you know, in the real world it's always more complicated."

Most of the findings from the BDFFP have indeed been variations on the theme of loss. Six species of primates can be found in the area of the project. Three of these—the black spider monkey, the brown capuchin monkey, and the bearded saki—are missing from the fragments. Birds like the long-tailed woodcreeper and the olive-backed foliage gleaner, which travel in mixed-species flocks, have all but disappeared from the smaller fragments and are found at much lower abundance in the larger ones. Frogs that breed in peccary wallows have vanished along with the peccaries that produced the wallows. Many species, sensitive even to slight changes in light and heat, have declined in abundance toward the edges of the fragments, though the number of light-loving butterflies has increased.

Meanwhile, though this is somewhat beyond the scope of the BDFFP, there's a dark synergy between fragmentation and global warming, just as there is between global warming and ocean acidification, and between global warming and invasive species, and between invasive species and fragmentation. A species that needs to migrate to keep up with rising temperatures, but is trapped in a forest fragment—even a very large fragment—is a species that isn't likely to make it. One of the defining features of the Anthropocene is that the world is changing in ways that compel species to move, and another is that it's changing in ways that create barriers—roads, clear-cuts, cities—that prevent them from doing so.

"The whole new layer on top of what I was thinking about in the nineteen-seventies is climate change," Lovejoy told me. He has written that "in the face of climatic change, even natural climatic change, human activity has created an obstacle course for the dispersal of biodiversity," the result of which could be "one of the greatest biotic crises of all time."

That night everyone went to sleep early. After what felt like a few minutes but might have been a few hours, I was woken by the most extraordinary racket. The sound seemed to be coming from nowhere and everywhere. It would rise to a crescendo, fall off, and then, just as I was starting to fall back to sleep, start up again. I knew it was the mating call of some kind of frog, and I got out of my hammock and grabbed a flashlight to take a look around. I couldn't find the source of the noise, but I did come across an insect with a bioluminescent stripe, which I would have liked to put in a jar, had there been any jars to put it in. The next morning, Cohn-Haft pointed out a pair of Manaus slender-legged tree frogs, locked in amplexus. The frogs were an orangey brown, with shovel-shaped faces. The male, clamped on the female's back, was about half her size. I recalled having read that amphibians in the Amazon lowlands, so far at least, seem largely to have escaped chytrid. Cohn-Haft, who, along with everyone else, had been kept up by the din, described the frog's call as a "prolonged groaning that explodes into a roar and ends in a chuckling laugh."

After several cups of coffee, we set out to watch the ant parade. Lovejoy had planned to come with us, but when he went to go put on a long-sleeved shirt, a spider that had taken up residence in it bit him on the hand. The spider looked relatively ordinary, but the bite was turning an angry red, and Lovejoy's hand was going numb. It was decided that he should stay at the camp.

"The ideal method is to let the ants come in around you," Cohn-Haft explained as we hiked along. "Then there's no way out; it's like painting yourself into a corner. And the ants will come up on you, and they'll bite your clothes. And you're in the middle of the action." In the distance, he heard a rufous-throated antbird making a sound somewhere between a tweet and a cackle. As the name suggests, rufous-throated antbirds are obligate ant-followers, so Cohn-Haft took this as a promising sign. However, a few minutes later, when we reached the spot where we'd seen their endless

A white-plumed antbird (*Pithys albifrons*).

column the day before, the ants were nowhere to be found. Cohn-Haft heard two other antbirds calling from the trees: a white-plumed antbird, which makes a high-pitched whistling noise; and a white-chinned woodcreeper, which has an upbeat, twittery song. They, too, seemed to be looking for the ants.

"They're as confused as we are," Cohn-Haft said. He speculated that the ants had been moving their bivouac and had now gone into what is known as their statary phase. During this phase, the ants stay more or less in one place to raise a new generation. The statary phase can last for up to three weeks, which helps explain one of the more puzzling discoveries to come out of the BDFFP: even forest fragments large enough to support colonies of army ants end up losing their antbirds. Obligate ant-followers need foraging ants to follow, and apparently in the fragments there just aren't enough colonies to insure that one will always be active. Here again, Cohn-Haft told me, was a demonstration of the rain-forest's logic. The antbirds are so good at doing "exactly what they do" that they're extremely sensitive to any change that makes their particular form of doing more difficult.

"When you find one thing that depends on something else that, in turn, depends on something else, the whole series of interactions depends on constancy," he said. I thought about this as we trudged back to camp. If Cohn-Haft was right, then in its crazy, circus-like complexity the ant-bird-butterfly parade was actually a figure for the Amazon's stability. Only in a place where the rules of the game remain fixed is there time for butterflies to evolve to feed on the shit of birds that evolved to follow ants. Yes, I was disappointed that we hadn't found the ants. But I figured I had nothing on the birds.

CHAPTER X

THE NEW PANGAEA

Myotis lucifugus

THE BEST TIME TO TAKE A BAT CENSUS IS THE DEAD OF WINTER. Bats are what are known as "true hibernators"; when the mercury drops, they begin looking for a place to settle down, or really upside down, since bats in torpor hang by their toes. In the north-eastern United States, the first bats to go into hibernation are usually the little browns. Sometime in late October or early November, they seek out a sheltered space, like a cave or a mineshaft, where conditions are likely to remain stable. The little browns are soon joined by the tricolored bats and then by the big browns and the small-footed bats. The body temperature of a hibernating bat drops by fifty or sixty degrees, often to right around freezing. Its heartbeat slows, its immune system shuts down, and the bat, dangling by its feet, falls into a state close to suspended animation. Counting hibernating bats demands a strong neck, a good headlamp, and a warm pair of socks.

In March 2007, some wildlife biologists from Albany, New York, went to conduct a bat census at a cave just west of the city.

This was a routine event, so routine that their supervisor, Al Hicks, stayed behind at the office. As soon as the biologists arrived at the cave, they pulled out their cell phones.

"They said, 'Holy shit, there's dead bats everywhere,' " Hicks, who works for New York State's Department of Environmental Conservation, would later recall. Hicks instructed them to bring some carcasses back to the office. He also asked the biologists to photograph any live bats they could find. When Hicks examined the photos, he saw that the animals looked as if they had been dunked, nose first, in talcum powder. This was something he had never encountered before, and he began e-mailing the photographs to all the bat specialists he could think of. None of them had ever seen anything like it either. Some of Hicks's counterparts in other states took a joking tone. What they wanted to know, they said, was what those bats in New York were snorting.

Spring arrived. Bats all across New York and New England awoke from their torpor and flew off. The white powder remained a mystery. "We were thinking, Oh, boy, we hope this just goes away," Hicks told me. "It was like the Bush administration. And, like the Bush administration, it just wouldn't go away." Instead, it spread. The following winter, the same white powdery substance was found on bats in thirty-three caves in four different states. Meanwhile, bats kept dying. In some hibernacula, populations plunged by more than ninety percent. In one cave in Vermont, thousands of corpses dropped from the ceiling and piled up on the ground, like snowdrifts.

The bat die-off continued the following winter, spreading to five more states. It continued the winter after that, in three additional states, and, although in many places there are hardly any bats left to kill off, it continues to this day. The white powder is now known to be a cold-loving fungus—what's known as a psychrophile—that was accidentally imported to the U.S., probably from Europe. When it was first isolated, the fungus, from the

A little brown bat (*Myotis lucifugus*) with white-nose syndrome.

genus *Geomyces*, had no name. For its effect on the bats it was dubbed *Geomyces destructans*.

WITHOUT human help, long-distance travel is for most species difficult, bordering on impossible. This fact was, to Darwin, central. His theory of descent with modification demanded that each species arise at a single place of origin. To spread from there, it either slithered or swam or loped or crawled or cast its seeds upon the wind. Given a long enough time, even a sedentary organism, like, say, a fungus, could, Darwin thought, become widely dispersed. But it was the limits of dispersal that made things interesting. These accounted for life's richness and, at the same time, for the patterns that could be discerned amid the variety. The barriers imposed by the oceans, for instance, explained why vast tracts of South America, Africa, and Australia, though in Darwin's words "entirely similar" in climate and topography, were populated by entirely *dis*similar flora and fauna. The creatures on each continent had evolved separately, and in this way, physical isolation had been transmuted into biological disparity. Similarly, the barriers imposed by land explained why the fish of the eastern Pacific were distinct from the fish of the western Caribbean, though these two groups were, as Darwin wrote, "separated only by the narrow, but impassable, isthmus of Panama." On a more local level, the species found on one side of a mountain range or a major river were often different from the species found on the other, though usually—and significantly—they were related. Thus, for example, Darwin noted, "the plains near the Straits of Magellan are inhabited by one species of Rhea, and northward the plains of La Plata by another species of the same genus, and not by a true ostrich or emu, like those found in Africa and Australia."

The limits of dispersal concerned Darwin in another way, too, this one harder to account for. As he'd seen firsthand, even remote volcanic islands, like the Galápagos, were full of life. Indeed, islands were home to many of the world's most marvelous creatures. For his theory of evolution to be correct, these creatures must be the descendants of colonizers. But how had the original colonizers arrived? In the case of the Galápagos, five hundred miles of open water separated the archipelago from the coast of South America. So vexed was Darwin by this problem that he spent over a year trying to replicate the conditions of an ocean crossing in the garden of his home in Kent. He collected seeds and immersed them in tanks of salt water. Every few days, he dredged out some of the seeds and planted them. The exercise proved time-consuming, for, he wrote to a friend, "the water I find must be renewed every other day, as it gets to smell horribly." But the results, he thought, were promising; barley seeds still germinated after four weeks' immersion, cress seeds after six, though the seeds "gave out a surprising quantity of slime." If an ocean current flowed at the rate of roughly one mile per hour, then over the course of six weeks a seed could be carried more than a thousand miles. How about an animal? Here Darwin's methods became even more baroque. He sliced off a pair of duck's feet and suspended them in a tank filled with snail hatchlings. After allowing the duck's feet to soak for a while, he lifted them out and had his children count how many hatchlings were attached. The tiny mollusks, Darwin found, could survive out of water for up to twenty hours, and in this length of time, he calculated, a duck with its feet attached might cover six or seven hundred miles. It was no mere coincidence, he observed, that many remote islands have no native mammals save for bats, which can fly.

Darwin's ideas about what he termed "geographical distribution" had profound implications, some of which would not be recognized until decades after his death. In the late nineteenth century, paleontologists began to catalog the many curious correspondences exhibited by fossils gathered on different continents. *Mesosaurus*, for example, is a skinny reptile with splayed-out teeth that lived during the Permian period. *Mesosaurus* remains turn up both in Africa and, an ocean away, in South America. *Glossopteris* is a tongue-shaped fern, also from the Permian period. Its fossils can be found in Africa, in South America, and in Australia. Since it was hard to see how a large reptile could have crossed the Atlantic, or a plant both the Atlantic and the Pacific, vast land bridges extending for several thousand miles were invoked. Why these ocean-spanning bridges had vanished and where they had gone to no one knew; presumably, they had sunk beneath the waves. In the early years of the twentieth century, the German meteorologist Alfred Wegener came up with a better idea.

"The continents must have shifted," he wrote. "South America must have lain alongside Africa and formed a unified block. . . . The two parts must then have become increasingly separated over a period of millions of years like pieces of a cracked ice floe in water." At one time, Wegener hypothesized, all of the present-day continents had formed one giant supercontinent, Pangaea. Wegener's theory of "continental drift," widely derided during his lifetime, was, of course, to a large extent vindicated by the discovery of plate tectonics.

One of the striking characteristics of the Anthropocene is the hash it's made of the principles of geographic distribution. If highways, clear-cuts, and soybean plantations create islands where none before existed, global trade and global travel do the reverse: they deny even the remotest islands their remoteness. The process of remixing the world's flora and fauna, which began slowly, along the routes of early human migration, has, in recent decades, accelerated to the point where in some parts of the world, non-native plants now outnumber native ones. During any given twenty-four-hour period, it is estimated that ten thousand different species are being moved around the world just in ballast water. Thus a single supertanker (or, for that matter, a jet passenger) can undo millions of years of geographic separation. Anthony Ricciardi, a specialist in introduced species at McGill University, has dubbed the current reshuffling of the earth's biota a "mass invasion event." It is, he has written, "without precedent" in the planet's history.

As it happens, I live just east of Albany, relatively close to the cave where the first piles of dead bats were discovered. By the time I learned about what was going on, white-nose syndrome, as it had become known, had spread as far as West Virginia and had killed something like a million bats. I called up Al Hicks, and he suggested since it was once again bat census season that I tag along for the next count. On a cold, gray morning we met up in a parking lot not far from his office. From there, we headed almost due north, toward the Adirondacks.

About two hours later, we arrived at the base of a mountain not far from Lake Champlain. In the nineteenth century and then again during World War II, the Adirondacks were a major source of iron ore, and shafts were sunk deep into the mountains. When the ore was gone, the shafts were abandoned by people and colonized by bats. For the census, we were going to enter a shaft of what was once the Barton Hill Mine. The entrance was halfway up the mountainside, which was covered in several feet of snow. At the trailhead, more than a dozen people were standing around stomping their feet against the cold. Most, like Hicks, worked for New York State, but there were also a couple of biologists from the U.S. Fish and Wildlife Service and a local novelist who was doing research for a book into which he was hoping to weave a white-nose subplot.

Everyone put on snowshoes, except for the novelist, who, it seemed, had missed the message to bring a pair. The snow was icy and the going slow, so it took half an hour to get maybe half a mile. While we were waiting for the novelist to catch up—he was having trouble with the three-foot-deep drifts—the conversation turned to the potential dangers of entering an abandoned mine. These, I was told, included getting crushed by falling rocks, being poisoned by a gas leak, and plunging over a sheer drop of a hundred feet or more. After another half an hour or so, we reached the mine entrance—essentially a large hole cut into the hillside. The stones in front of the entrance were white with bird droppings, and the snow was covered with paw prints. Evidently, ravens and coyotes had discovered that the spot was an easy place to pick up dinner.

"Well, shit," Hicks said. Bats were fluttering in and out of the mine, and in some cases crawling around on the snow. Hicks went to catch one; it was so lethargic that he grabbed it on the first try. He held it between his thumb and forefinger, snapped its neck, and placed it in a Ziploc bag. "Short survey today," he announced.

We unstrapped our snowshoes, put on helmets and headlamps, and filed into the mine, down a long, sloping tunnel. Shattered beams littered the ground, and bats flew up at us through the gloom. Hicks cautioned everyone to stay alert. "There's places that if you take a step you won't be stepping back," he warned. The tunnel twisted along, sometimes opening up into concert-hall-sized chambers with side tunnels leading out of them. Some of the chambers had acquired names; when we reached a sepulchral stretch known as the Don Thomas section, we split up into groups to start the survey. The process consisted of photographing as many bats as possible. (Later on, back in Albany, somebody sitting at a computer screen would have to count all the bats in the pictures.) I went with Hicks, who was carrying an enormous camera, and one of the biologists from the Fish and Wildlife Service, who had a laser pointer. Bats are highly social animals, and in the mine they hung from the rock ceiling in crowded clusters. Most were little brown bats—*Myotis lucifugus*, or "lucis" in bat-counting jargon. These are the dominant bat in the northeastern U.S. and the sort most likely to be seen fluttering around on a summer night. As the name suggests, they're little—only about five inches long and two-tenths of an ounce in weight—and brown, with lighter-colored fur on their bellies. (The poet Randall Jarrell described them as being "the color of coffee with cream in it.") Hanging from the ceiling, with their wings folded, they looked like damp pompoms. There were also small-footed bats (*Myotis leibii*), which can be identified by their very dark faces, and Indiana bats (*Myotis sodalis*), which, even before white-nose, were listed as an endangered species. As we moved along, we kept disturbing the bats, which squeaked and rustled around, like half-asleep children.

Despite the name, white-nose is not confined to bats' noses; as we worked our way deeper into the mine, people kept finding bats with freckles of fungus on their wings and ears. Several of these were dispatched, for study purposes, with a thumb and forefinger. Each dead bat was sexed—males can be identified by their tiny penises—and placed in a Ziploc bag.

Still today, it is not entirely understood how *Geomyces destructans* kills bats. What is known is that bats with white-nose often wake up from their torpor and fly around in the middle of the day. It's been hypothesized that the fungus, which, quite literally, eats away at the bats' skin, irritates the animals to the point of arousal. This, in turn, causes them to use up the fat stores that were supposed to take them through the winter. On the edge of starvation, they fly out into the open to search for insects, which, of course, at that time of year are not available. It's also been proposed that the fungus causes the bats to lose moisture through their skin. This leads them to become dehydrated, which prompts them to wake up to go in search of water. Again they use up critical energy stores and wind up emaciated and, finally, dead.

We had entered the Barton Hill Mine at around 1 PM. By 7 PM we were almost back where we'd started, at the bottom of the mountain, except that now we were on the inside of it. We came to a huge, rusty winch, which, when the mine was operational, had been used to haul ore to the surface. Below it, the path disappeared into a pool of water, black like the River Styx. It was impossible to go farther, and so we began the long climb up.

THE movement of species around the world is sometimes compared to Russian roulette. As in the high-stakes game, two very different things can happen when a new organism shows up. The first, which might be called the empty chamber option, is nothing. Either because the climate is unsuitable, or because the creature can't find enough to eat, or because it gets eaten itself, or for a host of other possible reasons, the new arrival doesn't survive (or at least fails to reproduce). Most potential introductions go unrecorded—indeed, entirely unheeded—so it's hard to get precise figures; almost certainly, though, the vast majority of potential invaders don't make it.

In the second option, not only does the introduced organism survive; it gives rise to a new generation, which in turn survives and gives rise to another generation. This is what's known in the invasive species community as "establishment." Again, it's impossible to say for sure how often this happens; many established species probably remain confined to the spot where they were introduced, or they're so innocuous they've gone unnoticed. But—and here's where the roulette analogy comes back—a certain number complete the third step in the invasion process, which is "spread." In 1916, a dozen strange beetles were discovered in a nursery near Riverton, New Jersey. By the following year, the insects, now known as *Popillia japonica* or, more commonly, as Japanese beetles, had dispersed in all directions and could be found over an area of three square miles. The year after that, the figure jumped to seven square miles and the year after that to forty-eight square miles. The beetle continued to expand its territory at a geometric rate, each year pushing out into a new concentric circle, and within two decades it could be found from Connecticut to Maryland. (It has since spread as far south as Alabama and as far west as Montana.) Roy van Driesche, an expert on invasive species at the University of Massachusetts, has estimated that out of every hundred potential introductions, somewhere between five and fifteen will succeed in establishing themselves. Of these five to fifteen, one will turn out to be the "bullet in the chamber."

Why some introduced species are able to proliferate explosively is a matter of debate. One possibility is that for species, as for grifters, there are advantages to remaining on the move. A species that's been transported to a new spot, especially on a new continent, has left many of its rivals and predators behind. This shaking free of foes, which is really the shaking free of evolutionary history, is referred to as "enemy release." There are lots of organisms that appear to have benefited from enemy release, including purple loosestrife, which arrived in the northeastern United States from Europe in the early nineteenth century. In its native habitat, purple loosestrife has all sorts of specialized enemies, including the black-margined loosestrife beetle, the golden loosestrife beetle, the loosestrife root weevil, and the loosestrife flower weevil. All of these were absent in North America when the plant appeared, which helps explain why it's been able to take over boggy areas from West Virginia to Washington State. Some of these specialized predators have recently been introduced into the U.S. in an effort to control the plant's spread. This sort of it-takes-an-invasive-to-catch-an-invasive strategy has a decidedly mixed record. In some cases it's proven highly successful; in other it's turned out to be another ecological disaster. To the latter category belongs the rosy wolfsnail—*Euglandina rosea*—which was introduced to Hawaii in the late nineteen-fifties. The wolfsnail, a native of Central America, was brought in to prey on a previously introduced species, the giant African snail—*Achatina fulica*—which had become an agricultural pest. *Euglandina rosea* mostly left *Achatina fulica* alone and focused its attention instead on Hawaii's small, colorful native snails. Of the more than seven hundred species of endemic snails that once inhabited the islands, something like ninety percent are now extinct, and those that remain are in steep decline.

The corollary to leaving old antagonists behind is finding new, naive organisms to take advantage of. A particularly famous—and ghastly—instance of this comes in the long, skinny form of the brown tree snake, *Boiga irregularis*. The snake is native to Papua New Guinea and northern Australia, and it found its way to Guam in the nineteen-forties, probably in military cargo. The only snake indigenous to the island is a small, sightless creature the size of a worm; thus Guam's fauna was entirely unprepared for *Boiga irregularis* and its voracious feeding habits. The snake ate its way through most of the island's native birds, including the Guam flycatcher, last seen in 1984; the Guam rail, which survives only owing to a captive breeding program; and the Mariana fruit-dove, which is extinct on Guam (though it persists on a couple of other, smaller islands). Before the tree snake arrived, Guam had three native species of mammals, all bats; today only one—the Marianas flying fox—remains, and it is considered highly endangered. Meanwhile, the snake, also a beneficiary of enemy release, was multiplying like crazy; at the peak of what is sometimes called its "irruption," population densities were as high as forty snakes per acre. So thorough has been the devastation wrought by the brown tree snake that it has practically run out of native animals to consume; nowadays it feeds mostly on other interlopers, like the curious skink, a lizard also introduced to Guam from Papua New Guinea. The author David Quammen cautions that while it is easy to demonize the brown tree snake, the animal is not evil; it's just amoral and in the wrong place. What *Boiga irregularis* has done in Guam, he observes, "is precisely what *Homo sapiens* has done all over the planet: succeeded extravagantly at the expense of other species."

With introduced pathogens, the situation is much the same. Long-term relationships between pathogens and their hosts are often characterized in military terms; the two are locked in an "evolutionary arms race," in which, to survive, each must prevent the other from getting too far ahead. When an entirely new pathogen shows up, it's like bringing a gun to a knife fight. Never having encountered the fungus (or virus or bacterium) before, the new host has no defenses against it. Such "novel interactions," as they're called, can be spectacularly deadly. In the eighteen hundreds, the American chestnut was the dominant deciduous tree in eastern forests; in places like Connecticut, it made up close to half the standing timber. (The tree, which can resprout from the roots, did fine even when heavily logged; "not only was baby's crib likely made of chestnut," a plant pathologist named George Hepting once wrote, "but chances were, so was the old man's coffin.") Then, around the turn of the century, *Cryphonectria parasitica*, the fungus responsible for chestnut blight, was imported to the U.S., probably from Japan. Asian chestnut trees, having coevolved with *Cryphonectria parasitica*, were easily able to withstand the fungus, but for the American species it proved almost a hundred percent lethal. By the nineteen-fifties, it had killed off practically every chestnut in the U.S.—some four billion trees. Several species of moths that depended on the tree disappeared along with it. Presumably it's the "novelty" of the chytrid fungus that accounts for its deadliness as well. It explains why, all of a sudden, golden frogs disappeared from Thousand Frog Stream and why amphibians in general are the planet's most threatened class of organism.

Even before the cause of white-nose syndrome was identified, Al Hicks and his colleagues suspected an introduced species. Whatever was killing the bats was presumably something they'd never encountered before, since the mortality rate was so high. Meanwhile, the syndrome was spreading from upstate New York in a classic bull's-eye pattern. This seemed to indicate that the killer had touched down near Albany. Suggestively, when the die-off began to make national news, a spelunker sent Hicks some photographs he'd shot about forty miles west of the city. The photos dated from 2006, a full year before Hicks's coworkers had called him to say "Holy shit," and they showed bats with clear signs of white-nose. The spelunker had taken his pictures in a cave connected to Howe Caverns, a popular tourist destination which offers, among other attractions, flashlight tours and underground boat trips.

"It's kind of interesting that the first record we have of this is photographs from a commercial cave in New York that gets about two hundred thousand visits a year," Hicks told me.

INTRODUCED species are now so much a part of so many landscapes that chances are if you glance out your window you will see some. From where I'm sitting, in western Massachusetts, I see grass, which someone at some point planted and which most definitely is not native to New England. (Almost all the grasses in American lawns come from somewhere else, including Kentucky bluegrass.) Since my lawn is not particularly well kept, I also see lots of dandelions, which came over from Europe and spread just about everywhere, and garlic mustard, also from Europe, and broadleaf plantains, yet another invader from Europe. (Plantains—*Plantago major*—seem to have arrived with the very first white settlers and were such a reliable sign of their presence that the Native Americans referred to them as "white men's footsteps.") If I get up from my desk and walk past the edge of the lawn, I can also find:

multiflora rose, a prickly invasive from Asia; Queen Anne's lace, another introduction from Europe; burdock, similarly from Europe; and oriental bittersweet, whose name speaks to its origins. According to a study of specimens in Massachusetts herbaria, nearly a third of all plant species documented in the state are "naturalized newcomers." If I dig down a few inches, I'll encounter earthworms, which are also newcomers. Before Europeans arrived, New England had no earthworms of its own; the region's worms had all been wiped out by the last glaciation, and even after ten thousand years of relative warmth, North America's native worms had yet to recolonize the area. Earthworms eat through leaf litter and in this way dramatically alter the makeup of forest soils. (Although earthworms are beloved by gardeners, recent research has linked their introduction to a decline in native salamanders in the Northeast.) As I write this, several new and potentially disastrous invaders appear to be in the process of spreading in Massachusetts. These include, in addition to *Geomyces destructans*: the Asian long-horned beetle, an import from China that feeds on a variety of hardwood trees; the emerald ash borer, also from Asia, whose larvae tunnel through and thereby kill ash trees; and the zebra mussel, a freshwater import from Eastern Europe that has the nasty habit of attaching itself to any available surface and consuming everything in the water column.

"Stop Aquatic Hitchhikers," declares a sign by a lake down the road from where I live. "Clean *all* recreational equipment." The sign shows a picture of a boat entirely coated in zebra mussels, as if someone had mistakenly applied mollusks instead of paint.

Wherever you are reading this, the story line is going to be roughly the same, and this goes not just for other parts of the United States but all around the world. DAISIE, a database of invasives in Europe, tracks more than twelve thousand species. APASD (the Asian-Pacific Alien Species Database), FISNA (the Forest Invasive Species Network for Africa), IBIS (the Island Biodiversity and Invasive Species Database), and NEMESIS (the

National Exotic Marine and Estuarine Species Information System) track thousands more. In Australia, the problem is so severe that from preschool on, children are enlisted in the control effort. The city council in Townsville, north of Brisbane, urges kids to conduct "regular hunts" for cane toads, which were purposefully, albeit disastrously, introduced in the nineteen-thirties to control sugarcane beetles. (Cane toads are poisonous, and trusting native species, like the northern quoll, eat them and die.) To dispose of the toads humanely, the council instructs children to "cool them in a fridge for 12 hours" and then place them "in a freezer for another 12 hours." A recent study of visitors to Antarctica found that in a single summer season, tourists and researchers brought with them more than seventy thousand seeds from other continents. Already one plant species, *Poa annua*, a grass from Europe, has established itself on Antarctica; since Antarctica has only two native plant species, this means that a third of its plants are now invaders.

From the standpoint of the world's biota, global travel represents a radically new phenomenon and, at the same time, a replay of the very old. The drifting apart of the continents that Wegener deduced from the fossil record is now being reversed—another way in which humans are running geologic history backward and at high speed. Think of it as a souped-up version of plate tectonics, minus the plates. By transporting Asian species to North America, and North American species to Australia, and Australian species to Africa, and European species to Antarctica, we are, in effect, reassembling the world into one enormous supercontinent—what biologists sometimes refer to as the New Pangaea.

AEOLUS Cave, which is set into a wooded hillside in Dorset, Vermont, is believed to be the largest bat hibernaculum in New England; it is estimated that before white-nose hit, nearly three hundred thousand bats—some from as far away as Ontario and Rhode Island—came there to spend the winter. A few weeks after I went with Hicks to the Barton Hill Mine, he invited me to accompany him to Aeolus. This trip had been organized by the Vermont Fish and Wildlife Department, and at the bottom of the hill, instead of strapping on snowshoes, we all piled onto snowmobiles. The trail zigged up the mountain in a series of long switchbacks. The temperature—about twenty-five degrees—was far too low for bats to be active, but when we parked near the entrance to the cave I could see bats fluttering around. The most senior of the Vermont officials, Scott Darling, announced that before going any farther, we'd all have to put on latex gloves and Tyvek suits. This seemed to me to be paranoid—a gesture out of the novelist's white-nose subplot; soon, however, I came to see the sense of it.

Aeolus was created by water flow over the course of thousands and thousands of years. To keep people out, the Nature Conservancy, which owns the cave, has blocked off the entrance with huge iron slats. With a key, one of the horizontal slats can be removed;

this creates a narrow gap that can be crawled (or slithered) through. Despite the cold, a sickening smell emanated from the opening—half game farm, half garbage dump. The stone path leading to the gate was icy and difficult to get a footing on. When it was my turn, I squeezed between the slats and immediately slid into something soft and dank. This, I realized, picking myself back up, was a pile of dead bats.

The entrance chamber of the cave, known as Guano Hall, is maybe thirty feet wide and twenty feet high at the front. Toward the back, it narrows and slopes. The tunnels that branch off from there are accessible only to spelunkers, and the tunnels that branch off from those are accessible only to bats. Peering into Guano Hall, I had the sense I was staring into a giant gullet. The scene, in the dimness, was horrific. There were long icicles hanging from the ceiling, and from the floor large knobs of ice rose up, like polyps. The ground was covered with dead bats; some of the ice knobs, I noticed, had bats frozen into them. There were torpid bats roosting on the ceiling, and also wide-awake ones, which would take off and fly by or, sometimes, right into us.

Why bat corpses pile up in some places, while in others they get eaten or in some other way disappear, is unclear. Hicks speculated that the conditions at Aeolus were so harsh that the bats didn't even make it out of the cave before dropping dead. He and Darling had planned to do a count of the bats in Guano Hall, but this plan was quickly abandoned in favor of just collecting specimens. Darling explained that the specimens would be going to the American Museum of Natural History, so that there would at least be a record of the hundreds of thousands of *lucis* and northern long-eared and tricolored bats that had once wintered in Aeolus. "This may be one of the last opportunities," he said. In contrast to a mine, which has been around for at most a few centuries, Aeolus, he pointed out, has existed for millennia. It's likely that bats have been hibernating there, generation after generation, since the cave's entrance was exposed at the end of the last ice age.

"That's what makes this so dramatic—it's breaking the evolutionary chain," Darling said. He and Hicks began picking dead bats off the ground. Those that were too badly decomposed were tossed back; those that were more or less intact were sexed and placed in two-quart plastic bags. I helped out by holding the bag for dead females. Soon it was full and another one was started. When the specimen count hit somewhere around five hundred, Darling decided that it was time to go. Hicks hung back; he'd brought along his enormous camera and said that he wanted to take more pictures. In the hours we had been slipping around in the cave, the carnage had grown even more grotesque; many of the bat carcasses had been crushed, and now there was blood oozing out of them. As I made my way up toward the entrance, Hicks called after me: "Don't step on any dead bats." It took me a moment to realize he was joking.

When, exactly, the New Pangaea project began is difficult to say. If you count people as an invasive species—the science writer Alan Burdick has called *Homo sapiens* "arguably the most successful invader in biological history"—the process goes back a hundred and twenty thousand years or so, to the period when modern humans first migrated out of Africa. By the time humans pushed into North America, around thirteen thousand years ago, they had domesticated dogs, which they brought with them across the Bering land bridge. The Polynesians who settled Hawaii around fifteen hundred years ago were accompanied not only by rats but also by lice, fleas, and pigs. The "discovery" of the New World initiated a vast biological swap meet—the so-called Columbian Exchange—which took the process to a whole new level. Even as Darwin was elaborating the principles of geographic distribution, those principles were being deliberately undermined by groups known as acclimatization societies. The very year *On the Origin of Species* was published, a member of an acclimatization society based in Melbourne released the first rabbits into Australia. They've been breeding there like, well, rabbits ever since. In 1890, a New York group that took as its mission "the introduction and acclimatization of such foreign varieties of the animal and vegetable kingdom as might prove useful or interesting" imported European starlings to the U.S. (The head of the group supposedly wanted to bring to America all the birds mentioned in Shakespeare.) A hundred starlings let loose in Central Park have by now multiplied to more than two hundred million.

Still today, Americans often deliberately import "foreign varieties" they think "might prove useful or interesting." Garden catalogs are filled with non-native plants, and aquarium catalogs with non-native fish. According to the entry on pets in the *Encyclopedia of Biological Invasions*, every year more non-indigenous species of mammals, birds, amphibians, turtles, lizards, and snakes are brought into the U.S. than the country has native species of these groups. Meanwhile, as the pace and volume of global trade have picked up, so, too, has the number of accidental imports. Species that couldn't survive an ocean crossing at the bottom of a canoe or in the hold of a whaling ship may easily withstand the same journey in the ballast tank of a modern cargo vessel or the bay of an airplane or in a tourist's suitcase. A recent study of non-indigenous species in North American coastal waters found that the "rate of reported invasions has increased exponentially over the past two hundred years." It attributed the accelerating pace to the increased quantities of goods being transported and also to the increased speed with which they travel. The Center for Invasive Species Research, which is based at the University of California-Riverside, estimates that California is now acquiring a new invasive species every sixty days. This is slow compared to Hawaii, where a new invader is added each month. (For comparison's sake, it's worth noting that before humans settled Hawaii, new species seem to have succeeded in establishing themselves on the archipelago roughly once every ten thousand years.)

The immediate effect of all this reshuffling is a rise in what might be called local diversity. Pick any place on earth—Australia, the Antarctic Peninsula, your local park—and, more likely than not, over the last few hundred years the number of species that can be found in the area has grown. Before humans arrived on the scene, many whole categories of organisms were missing from Hawaii; these included not only rodents but also amphibians, terrestrial reptiles, and ungulates. The islands had no ants, aphids, or mosquitoes. People have, in this sense, enriched Hawaii greatly. But Hawaii was, in its prehuman days, home to thousands of species that existed nowhere else on the planet, and many of these endemics are now gone or disappearing. The losses include, in addition to the several hundred species of land snails, dozens of species of birds and more than a hundred species of ferns and flowering plants. For the same reasons that local diversity has, as a general rule, been increasing, global diversity—the total number of different species that can be found worldwide—has dropped.

The study of invasives is often said to have begun with Charles Elton, a British biologist who published his seminal work, *The Ecology of Invasions by Animals and Plants,* in 1958. To explain the apparently paradoxical effects of moving species around, Elton used the analogy of a set of glass tanks. Imagine that each of the tanks is filled with a different solution of chemicals. Then imagine every tank connected to its neighbors by long, narrow tubes. If the taps to the tubes were left open for just a minute each day, the solutions would slowly start to diffuse. The chemicals would recombine. Some new compounds would form and some of the original compounds would drop out. "It might take quite a long time before the whole system came into equilibrium," Elton wrote. Eventually, though, all of the tanks would hold the same solution. The variety would have been eliminated, which was just what could be expected to happen by bringing long-isolated plants and animals into contact.

"If we look far enough ahead, the eventual state of the biological world will become not more complex, but simpler—and poorer," Elton wrote.

Since Elton's day, ecologists have tried to quantify the effects of total global homogenization by means of a thought experiment. The experiment starts with the compression of all the world's landmasses into a single megacontinent. The species-area relationship is then used to estimate how much variety such a landmass would support. The difference between this figure and the diversity of the world as it actually is represents the loss implied by complete interconnectedness. In the case of terrestrial mammals, the difference is sixty-six percent, which is to say that a single-continent world would be expected to contain only about a third as many mammalian species as currently exist. For land birds, it's just under fifty percent, meaning such a world would contain half as many bird species as the present one.

If we look even farther ahead than Elton did—millions of years farther—the biological world will, in all likelihood, become more complex again. Assuming that eventually travel and global commerce cease, the New Pangaea will, figuratively speaking, begin to break up. The continents will again separate, and islands will be re-isolated. And as this happens, new species will evolve and radiate from the invasives that have been dispersed around the world. Hawaii perhaps will get giant rats and Australia giant bunnies.

THE winter after I visited Aeolus with Al Hicks and Scott Darling, I went back with another group of wildlife biologists. The scene in the cave was very different this time around but no less macabre. Over the course of the year, the piles of bloody dead bats had almost completely decomposed, and all that was left was a carpet of delicate bones, each no thicker than a pine needle.

Ryan Smith, of the Vermont Fish and Wildlife Department, and Susi von Oettingen, of the U.S. Fish and Wildlife Service, were running the census this time around. They started with a cluster of bats hanging at the widest part of Guano Hall. On closer inspection, Smith noticed that most animals in the cluster were already dead, their tiny feet hooked to the rock in rigor mortis. But he thought he saw some living bats among the corpses. He called out the number to von Oettingen, who'd brought along a pencil and some index cards.

"Two *lucis*," Smith said.

"Two *lucis*," von Oettingen repeated, writing the number down.

Smith worked his way deeper into the cave. Von Oettingen called me over and gestured toward a crack in the rock face. Apparently at one point there had been dozens of bats hibernating inside it. Now there was just a layer of black muck studded with toothpick-sized bones. She recalled having seen, on an earlier visit to the cave, a live bat trying to nuzzle a group of dead ones. "It just broke my heart," she said.

Bats' sociability has turned out to be a great boon to *Geomyces destructans*. In winter, when they cluster, infected bats transfer the fungus to uninfected ones. Those that make it until spring then disperse, carrying the fungus with them. In this way, *Geomyces destructans* passes from bat to bat and cave to cave.

It took Smith and von Oettingen only about twenty minutes to census the nearly empty Guano Hall. When they were done, von Oettingen tallied up the figures on her cards: eighty-eight *lucis*, one northern long-eared bat, three tricolored bats, and twenty bats of indeterminate species. The total came to 112. This was about a thirtieth of the bats that used to be counted in the hall in a typical year. "You just can't keep up with that kind of mortality," von Oettingen told me as we wriggled out through the opening in the slats. She noted that *lucis* reproduce very slowly—females produce only one pup per year—so even if some bats ultimately prove resistant to white-nose, it was hard to see how populations could rebound.

Since that winter—the winter of 2010—*Geomyces destructans* has been traced to Europe, where it appears to be widespread. The continent has its own bat species, for example, the greater mouse-eared bat, which is found from Turkey to the Netherlands. Greater mouse-eared bats carry white-nose but don't seem to be bothered by it, which suggests that they and the fungus evolved in tandem.

Meanwhile, the situation in New England remains bleak. I went back to Aeolus for the count in the winter of 2011. Just thirty-five live bats were found in Guano Hall. I returned to the cave in 2012. After we'd hiked all the way up to the entrance, the biologist I was with decided it would be a mistake to go on: the risk of disturbing any bats that might be left outweighed the benefits of counting them. I hiked up again in the winter of 2013. By this point, according to the U.S. Fish and Wildlife Service, white-nose had spread to twenty-two states and five Canadian provinces and had killed more than six million bats. Although the temperature was below freezing, a bat flew up at me as I stood in front of the slats. I counted ten bats clinging to the rock face around the entrance; most of them had the desiccated look of little mummies. The Vermont Fish and Wildlife Department had posted signs on two trees near the entrance to Aeolus. One said: "This cave is closed until further notice." The other announced that violators could be fined "up to $1000 per bat." (It was unclear whether the sign referred to living animals or to the much more plentiful dead ones.)

The same corner of Guano Hall photographed in, from left to right: the winter of 2009 (with hibernating bats), the winter of 2010 (with fewer bats), and the winter of 2011 (with no bats).

Not long ago, I called Scott Darling to get an update. He told me that the little brown bat, once pretty much ubiquitous in Vermont, is now officially listed as an endangered species in the state. So, too, are northern long-eared and tricolored bats. "I frequently use the word 'desperate,'" he said. "We are in a desperate situation."

"As a brief aside," he went on, "I read this news story the other day. A place called the Vermont Center for Ecostudies has set up this Web site. People can take a photo of any and all organisms in Vermont and get them registered on this site. If I had read that a few years ago, I would have laughed. I would have said, 'You're going to have people sending in a picture of a *pine tree*?' And now, after what's happened with the little browns, I just wish they had done it earlier."

CHAPTER XI

THE RHINO GETS AN ULTRASOUND

Dicerorhinus sumatrensis

THE FIRST VIEW I GOT OF SUCI WAS HER PRODIGIOUS BACKSIDE. It was about three feet wide and stippled with coarse, reddish hair. Her ruddy brown skin had the texture of pebbled linoleum. Suci, a Sumatran rhino, lives at the Cincinnati Zoo, where she was born in 2004. The afternoon of my visit, several other people were also arrayed around her formidable rump. They were patting it affectionately, so I reached over and gave it a rub. It felt like petting a tree trunk.

Dr. Terri Roth, director of the zoo's Center for Conservation and Research of Endangered Wildlife, had arrived at the rhino's stall wearing scrubs. Roth is tall and thin, with long brown hair that she had pinned up in a bun. She pulled on a clear plastic glove that stretched over her right forearm, past the elbow, almost to her shoulder. One of Suci's keepers wrapped the rhino's tail in what looked like Saran Wrap and held it off to the side. Another keeper grabbed a pail and stationed himself by Suci's mouth. It was hard for me to see over Suci's bottom, but I was told he was feeding the rhino slices of apples, and I could hear her chomping away at them. While Suci was thus distracted, Roth pulled a second glove over the first and grabbed what looked like a video game remote. Then she stuck her arm into the rhino's anus.

Of the five species of rhinoceros that still exist, the Sumatran rhino—*Dicerorhinus sumatrensis*—is the smallest and, in a manner of speaking, the oldest. The genus *Dicerorhinus* arose some twenty million years ago, meaning that the Sumatran rhino's lineage goes back, relatively unchanged, to the Miocene. Genetic analysis has shown that the Sumatran is the closest living relative of the woolly rhino, which, during the last ice age, ranged from Scotland to South Korea. E. O. Wilson, who once spent an evening at the Cincinnati Zoo with Suci's mother and keeps a tuft of her hair on his desk, has described the Sumatran rhino as a "living fossil."

Sumatrans are shy, solitary creatures that in the wild seek out dense undergrowth. They have two horns—a large one at the tip of their snouts and a smaller one behind it—and pointy upper lips, which they use to grasp leaves and tree limbs. The animals' sex life is, from a human perspective at least, highly unpredictable. Females are what are known as induced ovulators; they won't release an egg unless they sense there's an eligible male around. In Suci's case, the nearest eligible male is ten thousand miles away, which is why Roth was standing there, with her arm up the rhino's rectum.

About a week earlier, Suci had been given a hormone injection designed to stimulate her ovaries. A few days after that, Roth had tried to artificially inseminate the rhino, a process that had involved threading a long, skinny tube through the folds of Suci's cervix, then pumping into it a vial of thawed semen. According to notes Roth had taken at the time, Suci had "behaved very well" during the procedure. Now it was time for a follow-up ultrasound. Grainy images appeared on a computer screen propped up near Roth's elbow. Roth located the rhino's bladder, which appeared on the screen as a dark bubble, then continued on. Her hope was that an egg in Suci's right ovary, which had been visible at the time of the insemination, had since been released. If it had, there was a chance Suci could become pregnant. But the egg was right where Roth had last seen it, a black circle in a cloud of gray.

"Suci did not ovulate," Roth announced to the half-dozen zookeepers who had gathered around to help. By this point, her entire arm had disappeared inside the rhino. The group let out a collective sigh. "Oh, no," someone said. Roth pulled out her arm and removed her gloves. Though clearly disappointed by the outcome, she didn't seem surprised by it.

THE Sumatran rhino was once found from the foothills of the Himalayas, in what's now Bhutan and northeastern India, down through Myanmar, Thailand, Cambodia, and the Malay Peninsula, and on the islands of Sumatra and Borneo. In the nineteenth century, it was still common enough that it was considered an agricultural pest. As southeast Asia's forests were felled, the rhino's habitat shrank and became fragmented. By the early nineteen-eighties, its population had been reduced to just a few hundred animals, most in isolated reserves on Sumatra and the rest in Malaysia. The animal seemed to be heading inexorably toward extinction when, in 1984, a group of conservationists gathered in Singapore to try to work out a rescue strategy. The plan they came up with called for, among other things, establishing a captive breeding program to insure against the species' total loss. Forty rhinos were caught, seven of which were sent to zoos in the U.S.

The captive breeding program got off to a disastrous start. Over a span of less than three weeks, five rhinos at a breeding facility in Peninsular Malaysia succumbed to trypanosomiasis, a disease caused by parasites spread by flies. Ten animals were caught in Sabah, a Malaysian state on the eastern tip of Borneo. Two of these died from injuries sustained during capture. A third was killed by tetanus. A fourth expired for unknown reasons, and, by the end of the decade, none had produced any offspring. In the U.S., the mortality rate was even higher. The zoos were feeding the animals hay, but, it turns out, Sumatran rhinos cannot live off hay; they require fresh leaves and branches. By the time anyone figured this out, only three of the seven animals that had been sent to America were still living, each in a different city. In 1995, the journal *Conservation Biology* published a paper on the captive breeding program. It was titled "Helping a Species Go Extinct."

That year, in a last-ditch effort, the Bronx and the Los Angeles Zoos sent their remaining rhinos—both females—to Cincinnati, which had the only surviving male, a bull named Ipuh. Roth was hired to figure out what to do with them. Being solitary, the animals couldn't be kept in the same enclosure, but obviously unless they were brought together, they couldn't mate. Roth threw herself into the study of rhino physiology, collecting blood samples, analyzing urine, and measuring hormone levels. The more she learned, the more the challenges multiplied.

"It's a very complicated species," she told me once we were back in her office, which is decorated with shelves full of wooden, clay, and plush rhinos. Rapunzel, the female from the Bronx, turned out to be too old to reproduce. Emi, the female from Los Angeles, seemed to be the right age but never seemed to ovulate, a puzzle that took Roth nearly a year to solve. Once she realized what the problem was—that the rhino needed to sense a male around—she began to arrange brief, carefully monitored "dates" between Emi and Ipuh. After a few months of fooling around, Emi got pregnant. Then she lost the pregnancy. She got pregnant again, and the same thing happened. This pattern kept repeating, for a total of five miscarriages. Both Emi and Ipuh developed eye problems, which Roth eventually determined were the result of too much time in the sun. (In the wild, Sumatran rhinos live in the shade of the forest canopy.) The Cincinnati Zoo invested a half a million dollars in custom-made awnings.

Emi got pregnant again in the fall of 2000. This time, Roth put

Suci at the Cincinnati Zoo.

her on liquid hormone supplements, which the rhino ingested in progesterone-soaked slices of bread. Finally, after a sixteen-month gestation, Emi gave birth to Andalas, a male. He was followed by Suci—the name means "sacred" in Indonesian—and then by another male, Harapan. In 2007, Andalas was shipped back to Sumatra, to a captive breeding facility in Way Kambas National Park. There, in 2012, he fathered a calf named Andatu—Emi and Ipuh's grandson.

The three captive-bred rhinos born in Cincinnati and the fourth in Way Kambas clearly don't make up for the many animals who died along the way. But they have turned out to be pretty much the only Sumatran rhinos born anywhere over the past three decades. Since the mid–nineteen-eighties, the number of Sumatran rhinos in the wild has declined precipitously, to the point that there are now believed to be fewer than a hundred left in the world. In an ironic twist, humans have brought the species so low that it seems only heroic human efforts can save it. If *Dicerorhinus sumatrensis* has a future, it's owing to Roth and the handful of others like her who know how to perform an ultrasound with one arm up a rhino's rectum.

And what's true of *Dicerorhinus sumatrensis* is, to one degree or another, true of all rhinos. The Javan rhino, which once ranged across most of southeast Asia, is now among the rarest animals on earth, with probably fewer than fifty individuals left, all in a single Javanese reserve. (The last known animal to exist somewhere else—in Vietnam—was killed by a poacher in the winter of 2010.) The Indian rhino, which is the largest of the five species and appears to be wearing a wrinkled coat, as in the Rudyard Kipling story, is down to around three thousand individuals, most living in four parks in the state of Assam. A hundred years ago, in Africa, the population of black rhinos approached a million; it has since been reduced to around five thousand animals. The white rhino, also from Africa, is the only rhino species not currently classified as threatened. It was hunted nearly to oblivion in the nineteenth century, made a remarkable comeback in the twentieth, and now, in the twenty-first, has come under renewed pressure from poachers, who can sell rhino horns on the black market for more than twenty thousand dollars a pound. (Rhino horns, which are made of keratin, like your fingernails, have long been used in traditional Chinese medicine but in recent years have become even more sought-after as a high-end party "drug"; at clubs in southeast Asia, powdered horn is snorted like cocaine.)

Meanwhile, of course, rhinos have plenty of company. People feel a deep, almost mystical sense of connection to big "charismatic" mammals, even if they're behind bars, which is why zoos devote so many resources to exhibiting rhinos and pandas and gorillas. (Wilson has described the evening he spent in Cincinnati with Emi as "one of the most memorable events" of his life.) But almost everywhere they're not locked up, big charismatic mammals are in trouble. Of the world's eight species of bears, six are categorized either as "vulnerable" to extinction or "endangered." Asian elephants have declined by fifty percent over the last three generations. African elephants are doing better, but, like rhinos, they're increasingly threatened by poaching. (A recent study concluded that the population of African forest elephants, which many consider to be a separate species from savanna elephants, has fallen by more than sixty percent just in the last ten years.) Most large cats—lions, tigers, cheetahs, jaguars—are in decline. A century from now, pandas and tigers and rhinos may well persist only in zoos or, as Tom Lovejoy has put it, in wildlife areas so small and heavily guarded they qualify as "quasi zoos."

THE day after Suci's ultrasound, I went to visit her again. It was a cold winter morning, and so Suci was confined to what is euphemistically referred to as her "barn"—a low-slung building made out of cinderblocks and filled with what look like prison cells. When I arrived, at around 7:30 AM, it was feeding time, and Suci was munching on some ficus leaves in one of the stalls. On an average day, the head rhino-keeper, Paul Reinhart, told me, she goes through about a hundred pounds of ficus, which has to be specially flown in from San Diego. (The total cost of the shipments comes to nearly a hundred thousand dollars a year.) She also consumes several gift baskets' worth of fruit; on this particular morning, the selection included apples, grapes, and bananas. Suci ate with what seemed to me to be lugubrious determination. Once the ficus leaves were gone, she started in on the branches. These were an inch or two thick, but she crunched through them easily, the way a person might bite through a pretzel.

Reinhart described Suci to me as a "good mix" between her mother, Emi, who died in 2009, and her father, Ipuh, who still lives at the Cincinnati Zoo. "Emi, if there was trouble to get into, she'd get into it," he recalled. "Suci, she's very playful. But she's also more hard-headed, like her dad." Another keeper walked by, pushing a large wheelbarrow full of steaming reddish-brown manure—Suci and Ipuh's output from the previous night.

Suci is so used to being around people, some of whom feed her treats and some of whom stick their hands up her rectum, that Reinhart let me hang out with her while he went off to do other chores. As I stroked her hairy flanks, I was reminded of an overgrown dog. (In fact, rhinos are most closely related to horses.) Though I can't say I sensed much playfulness, she did seem to me to be affectionate, and when I looked into her very black eyes, I could have sworn I saw a flicker of interspecies recognition. At the same time, I recalled the warning of one zoo official, who had told me that if Suci suddenly decided to jerk her enormous head, she could easily break my arm. After a while, it was time for the rhino to go get weighed. Some pieces of banana were laid out in front of a pallet scale built into the floor of the next stall over. When Suci trudged over to eat the bananas, the readout from the scale was 1,507 pounds.

Very big animals are, of course, very big for a reason. Already at birth, Suci weighed seventy pounds. Had she been born on Sumatra, at that point she could have fallen victim to a tiger (though nowadays Sumatran tigers, too, are critically endangered). But probably she would have been protected by her mother, and adult rhinos have no natural predators. The same goes for other so-called megaherbivores; full-grown elephants and hippos are so large that no animal dares attack them. Bears and big cats are similarly beyond predation.

Such are the advantages of being oversized—what might be called the "too big to quail" strategy—that it would seem, evolutionarily speaking, to be a pretty good gambit. And, indeed, at various points in its history, the earth has been full of colossal creatures. Toward the end of the Cretaceous, for instance, *Tyrannosaurus* was just one group of enormous dinosaurs; there was also the genus *Saltasaurus,* whose members weighed something like seven tons; *Therizinosaurus,* the largest of which were over thirty feet long; and *Saurolophus,* which were probably even longer.

Much more recently, toward the end of the last ice age, jumbo-sized animals could be found in pretty much all parts of the world. In addition to woolly rhinos and cave bears, Europe had aurochs, giant elk, and oversized hyenas. North America's behemoths included mastodons, mammoths, and *Camelops*, hefty cousins to modern camels. The continent was also home to: beavers the size of today's grizzlies; *Smilodon*, a group of saber-toothed cats; and *Megalonyx jeffersonii*, a ground sloth that weighed nearly a ton. South America had its own gigantic sloths, as well as *Toxodon*, a genus of mammals with rhino-like bodies and hippo-shaped heads, and glyptodonts, relatives of armadillos that, in some cases, grew to be as large as Fiat 500s. The strangest, most varied megafauna could be found in Australia. These included diprotodons, a group of lumbering marsupials colloquially known as rhinoceros wombats; *Thylacoleo carnifex*, a tiger-sized carnivore referred to as a marsupial lion; and the giant short-faced kangaroo, which reached a height of ten feet.

Even many relatively small islands had their own large beasts. Cyprus had a dwarf elephant and a dwarf hippopotamus. Madagascar was home to three species of pygmy hippos, a family of enormous flightless birds known as elephant birds, and several species of giant lemurs. New Zealand's megafauna was remarkable in that it was exclusively avian. The Australian paleontologist Tim Flannery has described it as a kind of thought experiment come to life: "It shows us what the world might have looked like if mammals as well as dinosaurs had become extinct 65 million years ago, leaving the birds to inherit the globe." On New Zealand, different species of moas evolved to fill the ecological niches occupied elsewhere by four-legged browsers like rhinos and deer. The largest of the moas, the North Island giant moa and the South Island giant moa, grew to be nearly twelve feet tall. Interestingly enough, the females were almost twice as giant as the giant males, and it is believed that the task of incubating the eggs fell to the fathers. New Zealand also had an enormous raptor, known as the Haast's

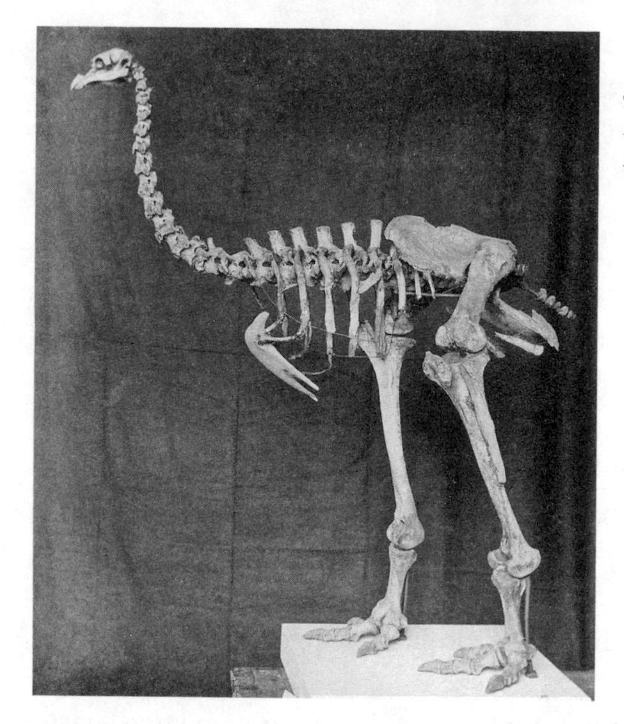

The largest moas grew to be nearly twelve feet tall.

eagle, which preyed on moas and had a wingspan of more than eight feet.

What happened to all these Brobdingnagian animals? Cuvier, who was the first to note their disappearance, believed they had been done in by the most recent catastrophe: a "revolution on the surface of the earth" that took place just before the start of recorded history. When later naturalists rejected Cuvier's catastrophism, they were left with a puzzle. Why *had* so many large beasts disappeared in such a relatively short amount of time?

"We live in a zoologically impoverished world, from which all the hugest, and fiercest, and strangest forms have recently disappeared," Alfred Russel Wallace observed. "And it is, no doubt, a much better world for us now they have gone. Yet it is surely a marvellous fact, and one that has hardly been sufficiently dwelt upon, this sudden dying out of so many large mammalia, not in one place only but over half the land surface of the globe."

* * *

As it happens, the Cincinnati Zoo is only about a forty-minute drive from Big Bone Lick, where Longueuil picked up the mastodon teeth that would inspire Cuvier's theory of extinction. Now a state park, Big Bone Lick advertises itself as the "birthplace of American vertebrate paleontology" and offers on its Web site a poem celebrating its place in history.

> At Big Bone Lick the first explorers
> found skeletons of elephants they said,
> found ribs of wooly mammoths, tusks.
>   The bones
> seemed wreckage from a mighty dream,
> a graveyard from a golden age.

One afternoon while visiting Suci, I decided to check out the park. The unmapped frontier of Longueuil's day is, of course, long gone, and the area is gradually being swallowed up by the Cincinnati suburbs. On the drive out, I passed the usual assortment of chain stores and then a series of housing developments, some so new the homes were still being framed. Eventually, I found myself in horse country. Just beyond the Woolly Mammoth Tree Farm, I turned into the park entrance. "No Hunting," the first sign said. Other signs pointed to a campsite, a lake, a gift shop, a minigolf course, a museum, and a herd of bison.

During the eighteenth and early nineteenth centuries, untold tons of specimens—mastodon femurs, mammoth tusks, giant ground sloth skulls—were hauled out of the bogs of Big Bone Lick. Some went to Paris and London, some to New York and Philadelphia. Still others were lost. (One whole shipment disappeared when a colonial trader was attacked by Kickapoo Indians; another sank on the Mississippi.) Thomas Jefferson proudly displayed bones from the Lick in an ad hoc museum he set up in the East Room of the White House. Lyell made a point of visiting the site during an American tour in 1842 and while there purchased for himself the teeth of a baby mastodon.

By now, Big Bone Lick has been so thoroughly picked over by collectors that there are hardly any big bones left. The park's paleontological museum consists of a single, mostly empty room. On one wall, there's a mural depicting a herd of melancholy-looking mammoths trudging across the tundra, and on the opposite wall some glass cases display a scattering of broken tusks and ground sloth vertebrae. Nearly as big as the museum is the adjacent gift shop, which sells wooden nickels and candy and T-shirts with the slogan, "I'm not fat—just big boned." A cheerful blonde was manning the shop's cash register when I visited. She told me that most people didn't appreciate the "significance of the park"; they just came for the lake and the minigolf, which, unfortunately, in winter was closed. Handing me a map, she urged me to follow the interpretive trail out back. I asked if she might be able to show me around, but she said, no, she was too busy. As far as I could tell, we were the only two people in the park.

I headed out along the trail. Just behind the museum, I came to a life-size mastodon, molded out of plastic. The mastodon had its head lowered, as if about to charge. Nearby was a ten-foot-tall plastic ground sloth, standing menacingly on its hind legs, and a mammoth that appeared to be sinking in terror into a bog. A dead, half-decomposed plastic bison, a plastic vulture, and some scattered plastic bones completed the grisly tableau.

Farther on, I came to Big Bone Creek, which was frozen over. Beneath the ice, the creek bubbled lazily along. A spur on the trail led to a wooden deck built over a patch of marsh. The water here was open. It smelled sulfurous and had a chalky white coating. A sign on the deck explained that during the Ordovician, ocean had covered the region. It was the accumulated salt from this ancient seabed that had drawn animals to drink at Big Bone Lick, and in many cases to die there. A second sign noted that among the remains found at the Lick were "those of at least eight species that became extinct around ten thousand years ago." As I continued along the trail, I came to still more signs. These gave an explanation—actually two different explanations—for the mystery of the missing megafauna. One sign offered the following account: "The change from coniferous to deciduous forest, or maybe the warming climate that brought about that change, caused the continent-wide disappearance of the Lick's extinct animals." Another sign put the blame elsewhere. "Within a thousand years after man arrived, the large mammals were gone," it said. "It seems likely that paleo-Indians played at least some role in their demise."

As early as the eighteen-forties, both explanations for the megafauna extinction had been proposed. Lyell was among those who favored the first account, as he put it, the "great modification in climate" that had occurred with the ice age. Darwin, as was his wont, sided with Lyell, though in this case somewhat reluctantly. "I cannot feel quite easy about the glacial period and the extinction of large mammals," he wrote. Wallace, for his part, initially also favored a climatic gloss. "There must have been some physical cause for this great change," he observed in 1876. "Such a cause exists in the great and recent physical change known as 'the Glacial Epoch.'" Then he had a change of heart. "Looking at the whole subject again," he observed in his last book, *The World of Life,* "I am convinced that . . . the rapidity of the extinction of so many large Mammalia is actually due to man's agency." The whole thing, he said, was really "very obvious."

Since Lyell, there's been a great deal of back and forth on the question, which has implications that extend far beyond paleobiology. If climate change drove the megafauna extinct, then this presents yet another reason to worry about what we are doing to global temperatures. If, on the other hand, people were to blame—and it seems increasingly likely that they were—then the import is almost more disturbing. It would mean that the current extinction

Diprotodon optatum was the largest marsupial ever.

event began all the way back in the middle of the last ice age. It would mean that man was a killer—to use the term of art an "overkiller"—pretty much right from the start.

THERE are several lines of evidence that argue in favor—or really against—humans. One of these is the event's timing. The megafauna extinction, it's now clear, did not take place all at once, as Lyell and Wallace believed it had. Rather, it occurred in pulses. The first pulse, about forty thousand years ago, took out Australia's giants. A second pulse hit North America and South America some twenty-five thousand years later. Madagascar's giant lemurs, pygmy hippos, and elephant birds survived all the way into the Middle Ages. New Zealand's moas made it as far as the Renaissance.

It's hard to see how such a sequence could be squared with a single climate change event. The sequence of the pulses and the sequence of human settlement, meanwhile, line up almost exactly. Archaeological evidence shows that people arrived first in Australia, about fifty thousand years ago. Only much later did they reach the Americas, and only many thousands of years after that did they make it to Madagascar and New Zealand.

"When the chronology of extinction is critically set against the chronology of human migrations," Paul Martin of the University of Arizona wrote in "Prehistoric Overkill," his seminal paper on the subject, "man's arrival emerges as the only reasonable answer" to the megafauna's disappearance.

In a similar vein, Jared Diamond has observed: "Personally, I can't fathom why Australia's giants should have survived innumerable droughts in their tens of millions of years of Australian history, and then have chosen to drop dead almost simultaneously (at least on a time scale of millions of years) precisely and just coincidentally when the first humans arrived."

In addition to the timing, there's strong physical evidence implicating humans. Some of this comes in the form of poop.

Megaherbivores generate mega amounts of shit, as is clear to anyone who's ever spent time standing behind a rhino. The ordure provides sustenance to fungi known as *Sporormiella*. *Sporormiella* spores are quite tiny—almost invisible to the naked eye—but extremely durable. They can still be identified in sediments that have been buried for tens of thousands of years. Lots of spores indicate lots of large herbivores chomping and pooping away; few or no spores suggest their absence.

A couple of years ago, a team of researchers analyzed a sediment core from a site known as Lynch's Crater, in northeastern Australia. They found that fifty thousand years ago, *Sporormiella* counts in the area were high. Then, rather abruptly around forty-one thousand years ago, *Sporormiella* counts dropped almost to zero. Following the crash, the landscape started to burn. (The evidence here was tiny grains of charcoal.) After that, the vegetation in the region shifted, from the sorts of plants you'd find in a rainforest toward more dry-adapted plants, like acacia.

If climate drove the megafauna to extinction, a shift in vegetation should *precede* a drop in *Sporormiella*: first the landscape would have changed, then the animals that depended on the original vegetation would have disappeared. But just the opposite had happened. The team concluded that the only explanation that fit the data was "overkill." *Sporormiella* counts dropped prior to changes in the landscape because the death of the megafauna *caused* the landscape to change. With no more large herbivores around to eat away at the forest, fuel built up, which led to more frequent and more intense fires. This, in turn, pushed the vegetation toward fire-tolerant species.

The megafauna extinction in Australia "couldn't have been driven by climate," Chris Johnson, an ecologist at the University of Tasmania and one of the lead authors on the core study, told me when I spoke to him on the phone from his office in Hobart. "I think we can say that categorically."

Even clearer is the evidence from New Zealand. When the Maori reached New Zealand, around the time of Dante, they found nine species of moa living on the North and South Islands. By the time European settlers arrived, in the early eighteen hundreds, not a single moa was to be seen. What remained were huge middens of moa bones, as well as the ruins of large outdoor ovens—leftovers of great, big bird barbecues. A recent study concluded that the moas were probably eliminated in a matter of decades. A phrase survives in Maori referring, obliquely, to the slaughter: *Kua ngaro i te ngaro o te moa*. Or "lost as the moa is lost."

THOSE researchers who persist in believing that climate change killed the megafauna say that the certainty of Martin, Diamond, and Johnson is misplaced. In their view, nothing has been proved about the event, "categorically" or otherwise, and everything in the preceding paragraphs is oversimplified. The dates of the extinctions are not clear-cut; they don't line up neatly with human migration; and, in any case, correlation is not causation. Perhaps most profoundly, they doubt the whole premise of ancient human deadliness. How could small bands of technologically primitive people have wiped out so many large, strong, and in some cases fierce animals over an area the size of Australia or North America?

John Alroy, an American paleobiologist who now works at Australia's Macquarie University, has spent a lot of time thinking about this question, which he considers a mathematical one. "A very large mammal is living on the edge with respect to its reproductive rate," he told me. "The gestation period of an elephant, for example, is twenty-two months. Elephants don't have twins, and they don't start to reproduce until they're in their teens. So these are big, big constraints on how fast they can reproduce, even if everything is going really well. And the reason they're able to exist at all is that when animals get to a certain size they escape from predation. They're no longer vulnerable to being attacked. It's a terrible strategy on the reproductive side, but it's a great advantage on the predator-avoidance side. And that advantage completely disappears when people show up. Because no matter how big an animal is, we don't have a constraint on what we can eat." This is another example of how a modus vivendi that worked for many millions of years can suddenly fail. Like the V-shaped graptolites or the ammonites or the dinosaurs, the megafauna weren't doing anything wrong; it's just that when humans appeared, "the rules of the survival game" changed.

Alroy has used computer simulations to test the "overkill" hypothesis. He's found that humans could have done in the megafauna with only modest effort. "If you've got one species that's providing what might be called a sustainable harvest, then other species can be allowed to go extinct without humans starving," he observed. For instance, in North America, white-tailed deer have a relatively high reproductive rate and therefore probably remained plentiful even as the number of mammoths dropped: "Mammoth became a luxury food, something you could enjoy once in a while, like a large truffle."

When Alroy ran the simulations for North America, he found

that even a very small initial population of humans—a hundred or so individuals—could, over the course of a millennium or two, multiply sufficiently to account for pretty much all of the extinctions in the record. This was the case even when the people were assumed to be only fair-to-middling hunters. All they had to do was pick off a mammoth or a giant ground sloth every so often, when the opportunity arose, and keep this up for several centuries. This would have been enough to drive the populations of slow-reproducing species first into decline and then, eventually, all the way down to zero. When Chris Johnson ran similar simulations for Australia, he came up with similar results: if every band of ten hunters killed off just one diprotodon a year, within about seven hundred years, every diprotodon within several hundred miles would have been gone. (Since different parts of Australia were probably hunted out at different times, Johnson estimates that continent-wide the extinction took a few thousand years.) From an earth history perspective, several hundred years or even several thousand is practically no time at all. From a human perspective, though, it's an immensity. For the people involved in it, the decline of the megafauna would have been so slow as to be imperceptible. They would have had no way of knowing that centuries earlier, mammoths and diprotodons had been much more common. Alroy has described the megafauna extinction as a "geologically instantaneous ecological catastrophe too gradual to be perceived by the people who unleashed it." It demonstrates, he has written, that humans "are capable of driving virtually any large mammal species extinct, even though they are also capable of going to great lengths to guarantee that they do not."

The Anthropocene is usually said to have begun with the industrial revolution, or perhaps even later, with the explosive growth in population that followed World War II. By this account, it's with the introduction of modern technologies—turbines, railroads, chainsaws—that humans became a world-altering force. But the megafauna extinction suggests otherwise. Before humans emerged on the scene, being large and slow to reproduce was a highly successful strategy, and outsized creatures dominated the planet. Then, in what amounts to a geologic instant, this strategy became a loser's game. And so it remains today, which is why elephants and bears and big cats are in so much trouble and why Suci is one of the world's last remaining Sumatran rhinos. Meanwhile, eliminating the megafauna didn't just eliminate the megafauna; in Australia at least it set off an ecological cascade that transformed the landscape. Though it might be nice to imagine there once was a time when man lived in harmony with nature, it's not clear that he ever really did.

CHAPTER XII

THE MADNESS GENE

Homo neanderthalensis

THE NEANDER VALLEY, OR, IN GERMAN, *DAS NEANDERTAL*, LIES about twenty miles north of Cologne, along a fold in the Düssel River, a sleepy tributary of the Rhine. For most of its existence, the valley was lined with limestone cliffs, and it was in a cave in the face of one of these cliffs that, in 1856, the bones were discovered that gave the world the Neanderthal. Today the valley is a sort of paleolithic theme park. In addition to the Neanderthal Museum, a strikingly modern building with walls of bottle green glass, there are cafés selling Neanderthal-brand beer, gardens planted with the sorts of shrubs that flourished during the ice ages, and hiking trails leading to the site of the find, though the bones, the cave, and even the cliffs are all gone. (The limestone was quarried and carted away as building blocks.) Directly inside the museum's entrance stands a model of an elderly Neanderthal smiling benignantly and leaning on a stick. He resembles an unkempt Yogi Berra. Next to him is one of the museum's most popular attractions: a booth called the Morphing Station. For three euros, visitors to the station can get a profile shot of themselves and, facing that, a second profile that's been doctored. In the doctored shot, the chin recedes, the forehead slopes, and the back of the head bulges out. Kids love to see themselves—or, better yet, their siblings—morphed into Neanderthals. They find it screamingly funny.

Since the discovery in the Neander Valley, Neanderthal bones have turned up all over Europe and the Middle East. They've been found as far north as Wales, as far south as Israel, and as far east as the Caucasus. Vast numbers of Neanderthal tools, too, have been unearthed. These include almond-shaped handaxes, knife-edged scrapers, and stone points that were probably hafted to spears. The tools were used to cut meat, to sharpen wood, and presumably also to prepare skins. The Neanderthals lived in Europe for at least a hundred thousand years. For the most part, this was a time of cold, and for stretches, it was intensely cold, with ice sheets covering Scandinavia. It is believed, though it's not known for certain, that, to protect themselves, the Neanderthals built shelters and fashioned some sort of clothing. Then, roughly thirty thousand years ago, the Neanderthals vanished.

All sorts of theories have been offered up to explain the vanishing. Often climate change is invoked, sometimes in the form of general instability leading up to what's referred to in earth science circles as the Last Glacial Maximum, and sometimes in the form of a "volcanic winter" that's believed to have been caused by an immense eruption not far from Ischia, in the area known as the Phlegraean Fields. Disease is also sometimes blamed, and so, too, is simple bad luck. In recent decades, though, it's become increasingly clear that the Neanderthal went the way of the *Megatherium*, the American mastodon, and the many other unfortunate megafauna. In other words, as one researcher put it to me, "their bad luck was us."

Modern humans arrived in Europe around forty thousand years ago, and again and again, the archaeological record shows, as soon as they made their way to a region where Neanderthals

were living, the Neanderthals in that region disappeared. Perhaps the Neanderthals were actively pursued, or perhaps they were just outcompeted. Either way, their decline fits the familiar pattern, with one important (and unsettling) difference. Before humans finally did in the Neanderthals, they had sex with them. As a result of this interaction, most people alive today are slightly—up to four percent—Neanderthal. A T-shirt available for sale near the Morphing Station puts the most upbeat spin possible on this inheritance. *ICH BIN STOLZ, EIN NEANDERTHALER ZU SEIN,* it declares in block capital letters. ("I am proud to be a Neanderthal.") I liked the T-shirt so much I bought one for my husband, though recently I realized that I've very rarely seen him wear it.

THE Max Planck Institute for Evolutionary Anthropology is situated almost three hundred miles due east of the Neander Valley, in the city of Leipzig. The institute occupies a spanking new building shaped a bit like a banana, and it stands out conspicuously in a neighborhood that still bears the stamp of the city's East German past. Just to the north is a block of Soviet-style apartment buildings. To the south stands a huge hall with a golden steeple, which used to be known as the Soviet Pavilion (and which is now empty). In the lobby of the institute there's a cafeteria and an exhibit on great apes. A TV in the cafeteria plays a live feed of the orangutans at the Leipzig Zoo.

Svante Pääbo heads the institute's department of evolutionary genetics. He is tall and lanky, with a long face, a narrow chin, and bushy eyebrows, which he often raises to emphasize some sort of irony. Pääbo's office is dominated by two figures. One of these is of Pääbo himself—a larger-than-life-size portrait that his graduate students presented to him on his fiftieth birthday. (Each student painted a piece of the portrait, the overall effect of which is a surprisingly good likeness, but in mismatched colors that make it look as if he has a skin disease.) The other figure is a Neanderthal—a life-size model skeleton, propped up so that its feet dangle over the floor.

Pääbo, who is Swedish, is sometimes called the "father of paleogenetics." He more or less invented the study of ancient DNA. His early work, as a graduate student, involved trying to extract genetic information from the flesh of Egyptian mummies. (He wanted to know who among the pharaohs was related to whom.) Later, he turned his attention to Tasmanian tigers and to giant ground sloths. He extracted DNA from the bones of mammoths and moas. All of these projects were groundbreaking at the time, yet all could be seen as just warm-up exercises for Pääbo's current, most extravagantly ambitious endeavor: sequencing the entire Neanderthal genome.

Pääbo announced the project in 2006, just in time for the 150th anniversary of the original Neanderthal's discovery. By then, a complete version of the human genome had already been published. So, too, had versions of the chimpanzee, mouse, and rat genomes. But humans, chimps, mice, and rats are, of course, living organisms. Sequencing the dead is a whole lot more difficult. When an organism expires, its genetic material begins to break down, so that instead of long strands of DNA, what's left, under the best of circumstances, are fragments. Trying to figure out how all the fragments fit together might be compared to trying to reassemble a Manhattan telephone book from pages that have been put through a shredder, mixed with yesterday's trash, and left to rot in a landfill.

When the project is completed, it should be possible to lay the human genome and the Neanderthal genome side by side and identify, base pair by base pair, exactly where they diverge. Neanderthals were extremely similar to modern humans; probably they were our very closest relatives. And yet clearly they were *not* humans. Somewhere in our DNA must lie the key mutation (or, more probably, mutations) that set us apart—the mutations that make us the sort of creature that could wipe out its nearest relative, then dig up its bones and reassemble its genome.

"I want to know what changed in fully modern humans, compared with Neanderthals, that made a difference," Pääbo told me. "What made it possible for us to build up these enormous societies, and spread around the globe, and develop the technology that I think no one can doubt is unique to humans? There has to be a genetic basis for that, and it is hiding somewhere in these lists."

THE bones from the Neanderthal Valley were discovered by quarry workers who treated them as rubbish. It's likely that they would have been lost entirely had the quarry's owner not heard about the find and insisted that the remains—a skullcap, a clavicle, four arm bones, two thighbones, parts of five ribs, and half a pelvis—be salvaged. Believing the bones to belong to a cave bear, the quarry owner passed them on to a local schoolteacher, Johann Carl Fuhlrott, who moonlighted as a fossilist. Fuhlrott realized that he was dealing with something at once stranger and more familiar than a bear. He declared the remains to be traces of a "primitive member of our race."

As it happened, this was right around the time that Darwin published *On the Origin of Species*, and the bones quickly got caught up in the debate over the origin of humans. Opponents of evolution dismissed Fuhlrott's claims. The bones, they said, belonged to an ordinary person. One theory held that it was a Cossack who had wandered into the region in the tumult following the Napoleonic Wars. The reason the bones looked odd—Neanderthal femurs are distinctly bowed—was that the Cossack had spent too long on his horse. Another attributed the remains to a man with rickets: the man had been in so much pain from his disease that he'd kept his forehead perpetually tensed—hence the protruding browridge. (What a man with rickets and in constant pain was doing climbing up a cliff and into a cave was never really explained.)

Over the next few decades, more bones like the ones from the Neander Valley—thicker than those of modern humans and with strangely shaped skulls—kept turning up. Clearly, all these finds could not be explained by tales of disoriented Cossacks or rickety spelunkers. But evolutionists, too, found the bones perplexing. Neanderthals had very large skulls—larger, on average, than people today. This made it hard to fit them into a narrative that started with small-brained apes and led progressively up to big-brained Victorians. In *The Descent of Man*, which appeared in 1871, Darwin alludes to Neanderthals only in passing. "It must be admitted that some skulls of very high antiquity, such as the famous one of Neanderthal, are well developed and capacious," he notes.

At once human and not, Neanderthals represent an obvious foil for ourselves, and a great deal that's been written about them since *The Descent of Man* reflects the awkwardness of this relationship. In 1908, a nearly complete skeleton was discovered in a cave near La Chapelle-aux-Saints, in southern France. It found its way to a paleontologist named Marcellin Boule, at Paris's Museum of Natural History. In a series of monographs, Boule invented what might be called the "don't-be-such-a Neanderthal" version of the Neanderthals: bent-kneed, hunched over, and brutish. Neanderthal bones, Boule wrote, displayed a "distinctly simian arrangement," while the shape of their skulls indicated "the predominance of functions of a purcly vegetative or bestial kind." Inventiveness, "artistic and religious sensibilities," and capacities for abstract thought were, according to Boule, clearly beyond such a beetle-browed creature. Boule's conclusions were studied and then echoed by many of his contemporaries; Sir Grafton Elliot Smith, a British anthropologist, for instance, described Neanderthals as walking with "a half-stooping slouch" upon "legs of a peculiarly ungraceful form." (Smith also claimed that Neanderthals' "unattractiveness" was "further emphasized by a shaggy covering of hair over

A Neanderthal as depicted in 1909.

most of the body," although there was—and still is—no physical evidence to suggest that they were hairy.)

In the nineteen-fifties, a pair of anatomists, William Straus and Alexander Cave, decided to reexamine the skeleton from La Chapelle. World War II—not to mention World War I—had shown the sort of brutishness the most modern of modern humans were capable of, and Neanderthals were due for a reappraisal. What Boule had taken for the Neanderthal's natural posture, Straus and Cave determined, was probably a function of arthritis. Neanderthals did not walk with a slouch or with bent knees. Indeed, given a shave and a new suit, the pair wrote, a Neanderthal probably would attract no more attention on a New York City subway "than some of its other denizens." More recent scholarship has tended to support the idea that Neanderthals, if not necessarily up to riding incognito on the IRT, certainly walked upright, with a gait we would recognize more or less as our own.

In the nineteen-sixties, an American archaeologist named Ralph Solecki uncovered the remains of several Neanderthals in a cave in northern Iraq. One of them, known as Shanidar I, or Nandy for short, had suffered a grievous head injury that had probably left him at least partially blind. His injuries had healed, which suggested that he must have been cared for by other members of his social group. Another, Shanidar IV, had apparently been buried, and the results of a soil analysis from the gravesite convinced Solecki that Shanidar IV had been interred with flowers. This he took as evidence of a deep Neanderthal spirituality.

"We are brought suddenly to the realization that the universality of mankind and the love of beauty go beyond the boundary of our own species," he wrote in a book about his discovery, *Shanidar: The First Flower People*. Some of Solecki's conclusions have since been challenged—it seems more likely that the flowers were brought into the cave by burrowing rodents than by grieving relatives—but his ideas had a wide influence, and it is Solecki's soulful near-humans who are on display in the Neander Valley. In the museum's dioramas,

A Neanderthal who's been given a shave and a new suit.

Neanderthals live in tepees, wear what look like leather yoga pants, and gaze contemplatively over the frozen landscape. "Neanderthal man was not some prehistoric Rambo," one of the display tags admonishes. "He was an intelligent individual."

DNA is often compared to a text, a comparison that's apt as long as the definition of "text" encompasses writing that doesn't make sense. DNA consists of molecules known as nucleotides knit together in the shape of a ladder—the famous double helix. Each nucleotide contains one of four bases: adenine, thymine, guanine, and cytosine, which are designated by the letters *A, T, G,* and *C,* so that a stretch of the human genome might be represented as *ACCTCCTCTAATGTCA*. (This is an actual sequence, from chromosome 10; the comparable sequence in an elephant is *ACCTCCCCTAATGTCA*.) The human genome is three billion bases—or, really, base pairs—long. As far as can be determined, most of it codes for nothing.

The process that turns an organism's long strands of DNA into fragments—from a "text" into something more like confetti—starts pretty much as soon as the organism expires. A good deal of the destruction is accomplished in the first few hours after death, by enzymes inside the creature's own body. After a while, all that remains are snippets, and after a longer while—how long seems to depend on the conditions of decomposition—these snippets, too, disintegrate. Once that happens, there's nothing for even the most dogged paleogeneticist to work with. "Maybe in the permafrost you could go back five hundred thousand years," Pääbo told me. "But it's certainly on this side of a million." Five hundred thousand years ago, the dinosaurs had been dead for about sixty-five million years, so the whole *Jurassic Park* fantasy is, sadly, just that. On the other hand, five hundred thousand years ago modern humans did not yet exist.

For the genome project, Pääbo managed to obtain twenty-one Neanderthal bones that had been found in a cave in Croatia. (In order to extract DNA, Pääbo, or any other paleogeneticist, has to cut up samples of bone and then dissolve them, a process that, for fairly obvious reasons, museums and fossil collectors are hesitant to sanction.) Only three of these bones yielded Neanderthal DNA. To compound the problem, that DNA was swamped by the DNA of microbes that had been feasting on the bones for the last thirty thousand years, which meant that most of the sequencing effort was going to waste. "There were times when one despaired," Pääbo told me. No sooner would one difficulty be solved than another would materialize. "It was an emotional roller coaster," recalled Ed Green, a biomolecular engineer from the University of California-Santa Cruz, who worked on the project for several years.

The project was finally generating useful results—essentially, long lists of *A*'s, *T*'s, *G*'s, and *C*'s—when one of the members of Pääbo's team, David Reich, a geneticist at Harvard Medical School, noticed something odd. The Neanderthal sequences were, as expected, very similar to human sequences. But they were more similar to some humans than to others. Specifically, Europeans and Asians shared more DNA with Neanderthals than did Africans. "We tried to make this result go away," Reich told me. "We thought, 'This must be wrong.' "

For the past twenty-five years or so, the study of human evolution has been dominated by the theory known in the popular press as "Out of Africa" and in academic circles as the "recent single-origin" or "replacement" hypothesis. This theory holds that all modern humans are descended from a small population that lived in Africa roughly two hundred thousand years ago. Around a hundred and twenty thousand years ago, a subset of that population migrated into the Middle East, and from there, further subsets eventually pushed northwest in Europe, east into Asia, and all the way east to Australia. As they moved north and east, modern humans encountered Neanderthals and other so-called archaic humans, who already inhabited those regions. The modern humans "replaced" the archaic humans, which is a nice way of saying they drove them to extinction. This model of migration and "replacement" implies that the relationship between Neanderthals and humans should be the same for all people alive today, regardless of where they come from.

Many members of Pääbo's team suspected that the Eurasian bias was a sign of contamination. At various points, the samples had been handled by Europeans and Asians; perhaps these people had got their DNA mixed in with the Neanderthals'. Several tests were run to assess this possibility. The results were all negative. "We kept seeing this pattern, and the more data we got, the more statistically overwhelming it became," Reich said. Gradually, the other team members started to come around. In a paper published in *Science* in May 2010, they introduced what Pääbo has come to refer to as the "leaky replacement" hypothesis. (The paper was later voted the journal's outstanding article of the year, and the team received a twenty-five-thousand-dollar prize.) Before modern humans "replaced" the Neanderthals, they had sex with them. The liaisons produced children, who helped to populate Europe, Asia, and the New World.

The leaky-replacement hypothesis—assuming for the moment that it's correct—provides the strongest possible evidence for the closeness of Neanderthals and modern humans. The two may or may not have fallen in love; still, they made love. Their hybrid children may or may not have been regarded as monsters; nevertheless someone—perhaps Neanderthals at first, perhaps humans—cared for them. Some of these hybrids survived to have kids of their own, who, in turn, had kids, and so on up to the present day. Even now, at least thirty thousand years after the fact, the signal is discernible: all non-Africans, from the New Guineans to the French to the Han Chinese, carry somewhere between one and four percent Neanderthal DNA.

One of Pääbo's favorite words in English is "cool." When he finally came around to the idea that Neanderthals bequeathed some of their genes to modern humans, he told me, "I thought it was very cool. It means that they are not totally extinct—that they live on a little bit in us."

THE Leipzig Zoo lies on the opposite side of the city from the Institute for Evolutionary Anthropology, but the institute has its own lab building on the grounds, as well as specially designed testing rooms inside the ape house, which is known as Pongoland. Since none of our very closest relatives survive (except as little bits in us), researchers have to rely on our next closest kin, chimpanzees and bonobos, and our somewhat more distant relations, gorillas and orangutans, to perform live experiments. (The same or, at least, analogous experiments are usually also performed on small children, to see how they compare.) One morning I went to the zoo, hoping to watch an experiment in progress. That day, a BBC crew was also visiting Pongoland, to film a program on animal intelligence, and when I arrived at the ape house I found it strewn with camera cases marked ANIMAL EINSTEINS.

For the benefit of the cameras, a researcher named Héctor Marín Manrique was reenacting a series of experiments he'd performed earlier in a more purely scientific spirit. A female orangutan named Dokana was led into one of the testing rooms. Like most orangutans, she had copper-colored fur and a world-weary expression. In the first experiment, which involved red juice and skinny tubes of plastic, Dokana showed that she could distinguish a functional drinking straw from a non-functional one. In the second, which involved more red juice and more plastic, she showed that she understood the *idea* of a straw by extracting a solid rod from a length of piping and using the now-empty pipe to drink through. Finally, in a Mensa-level display of pongid ingenuity, Dokana managed to get at a peanut that Manrique had placed at the bottom of a long plastic cylinder. (The cylinder was fixed to the wall, so it couldn't be knocked over.) She fist-walked over to her drinking water, took some water in her mouth, fist-walked back, and spat into the cylinder. She repeated the process until the peanut floated within reach. Later, I watched the BBC crew restage this experiment with some five-year-old children, using little plastic containers of candy in place of peanuts. Even though a full watering can had been left conspicuously nearby, only one of the kids—a girl—managed to work her way to the floating option, and this was after a great deal of prompting. ("How would water help me?" one of the boys asked querulously, just before giving up.)

One way to try to answer the question "What makes us human?" is to ask "What makes us different from great apes?" or, to be more precise, from nonhuman apes, since, of course, humans *are* apes. As just about every human by now knows—and as the experiments with Dokana once again confirm—nonhuman apes are extremely clever. They're capable of making inferences, of solving complex puzzles, and of understanding what other apes are (and are not) likely to know. When researchers from Leipzig performed a battery of tests on chimpanzees, orangutans, and two-and-a-half-year-old children, they found that the chimps, the orangutans, and the kids performed comparably on a wide range of tasks that involved understanding of the physical world. For example, if an experimenter placed a reward inside one of three cups, and then moved the cups around, the apes found the goody just as often as the kids—indeed, in the case of chimps, more often. The apes seemed to grasp quantity as well as the kids did—they consistently chose the dish containing more treats, even when the choice involved using what might loosely be called math—and also seemed to have just as good a grasp of causality. (The apes, for instance, understood that a cup that rattled when shaken was more likely to contain food than one that did not.) And they were equally skillful at manipulating simple tools.

Where the kids routinely outscored the apes was in tasks that involved reading social cues. When the children were given a hint about where to find a reward—someone pointing to or looking at the right container—they took it. The apes either didn't understand that they were being offered help or couldn't follow the cue. Similarly, when the children were shown how to obtain a reward, by, say, ripping open a box, they had no trouble grasping the point and imitating the behavior. The apes, once again, were flummoxed. Admittedly, the kids had a big advantage in the social realm, since the experimenters belonged to their own species. But, in general, apes seem to lack the impulse toward collective problem-solving that's so central to human society.

"Chimps do a lot of incredibly smart things," Michael Tomasello, who heads the institute's department of developmental and comparative psychology, told me. "But the main difference we've seen is 'putting our heads together.' If you were at the zoo today, you would never have seen two chimps carry something heavy together. They don't have this kind of collaborative project."

* * *

PÄÄBO usually works late, and most nights he has dinner at the institute, where the cafeteria stays open until 7 PM. One evening, though, he offered to knock off early and show me around downtown Leipzig. We visited the church where Bach is buried and ended up at Auerbachs Keller, the bar to which Mephistopheles brings Faust in the fifth scene of Goethe's play. (The bar was supposedly Goethe's favorite hangout when he was a university student.) I had been to the zoo the day before, and I asked Pääbo about a hypothetical experiment. If he had the opportunity to submit Neanderthals to the sorts of tests I'd seen in Pongoland, what would he do? What did he think they were like? Did he think they'd be able to talk? He sat back in his chair and folded his arms across his chest.

"One is so tempted to speculate," he said. "So I try to resist it by refusing questions such as 'Do I think they would have spoken?' Because, honestly, I don't know, and in some sense you can speculate with just as much justification as I can."

The many sites where their remains have been found give plenty of hints about what Neanderthals were like, at least to those inclined to speculate. Neanderthals were extremely tough—this is attested to by the thickness of their bones—and were probably capable of beating modern humans to a pulp. They were adept at making stone tools, though they seem to have spent tens of thousands of years making the same tools over and over again. At least on some occasions, they buried their dead. Also on some occasions, they appear to have killed and eaten each other. Not just Nandy but many Neanderthal skeletons show signs of disease or disfigurement. The original Neander Valley Neanderthal seems to have suffered from two serious injuries, one to his head and the other to his left arm. The La Chapelle Neanderthal endured, in addition to arthritis, a broken rib and kneecap. These injuries may reflect the rigors of hunting with the Neanderthals' limited repertoire of weapons; the Neanderthals never seem to have developed projectiles, so they would have to have gotten more or less on top of their prey in order to kill them. Like Nandy, both the original and the La Chapelle Neanderthal recovered from their injuries, which means that Neanderthals must have watched out for one another, which, in turn, implies a capacity for empathy. From the archaeological record, it's inferred that Neanderthals evolved in Europe or in western Asia and dispersed from there, stopping when they reached water or some other significant obstacle. (During the last glaciation, when sea levels were so much lower than they are now, there was no English Channel to contend with.) This is one of the most basic ways modern humans differ from Neanderthals, and, in Pääbo's view, it's also one of the most intriguing. When modern humans journeyed to Australia, even though it was the middle of an ice age, there was no way to make the trip without crossing open water.

Archaic humans like *Homo erectus* "spread like many other mammals in the Old World," Pääbo told me. "They never came to Madagascar, never to Australia. Neither did Neanderthals. It's only fully modern humans who start this thing of venturing out on the ocean where you don't see land. Part of that is technology, of course; you have to have ships to do it. But there is also, I like to think or say, some madness there. You know? How many people must have sailed out and vanished on the Pacific before you found Easter Island? I mean, it's ridiculous. And why do you do that? Is it for the glory? For immortality? For curiosity? And now we go to Mars. We never stop."

If Faustian restlessness is one of the defining characteristics of modern humans, then, by Pääbo's account, there must be some sort of Faustian gene. Several times, he told me that he thought it should be possible to identify the basis for our "madness" by comparing Neanderthal and human DNA. "If we one day will know that some freak mutation made the human insanity and exploration thing possible, it will be amazing to think that it was this little inversion on this chromosome that made all this happen and human

TACACTCACATTTTTTTGCATATTATCTAGTCCCATGACATTA

Neanderthal

TACACTCACATTTTTTTACATATTATCTAGCCCCATGACATTA

chimp

TACACTCACA-TTTTTTACATATTATCTAGTCCCATGACATTA

The same stretch of chromosome 5 from the human, Neanderthal, and chimp genomes.

changed the whole ecosystem of the planet and made us dominate everything," he said at one point. At another, he said, "We are crazy in some way. What drives it? That I would really like to understand. That would be really, really cool to know."

ONE afternoon, when I wandered into his office, Pääbo showed me a photograph of a skullcap that had recently been discovered by an amateur fossil collector about half an hour from Leipzig. From the photograph, which had been emailed to him, Pääbo had decided that the skullcap could be quite ancient. He thought it might belong to an early Neanderthal or even a *Homo heidelbergensis*, which some believe to be the common ancestor from which both humans and Neanderthals are descended. He'd also decided that he had to have it. The skullcap had been found at a quarry in a pool of water; perhaps, he theorized, these conditions had preserved it, so that if he got to it soon, he'd be able to extract some DNA. But the skull had already been promised to a professor of anthropology in Mainz. How could he persuade the professor to give him enough bone to test?

Pääbo called everyone he knew who he thought might know the professor. He had his secretary contact the professor's secretary to get the professor's private cell phone number, and joked—or maybe only half joked—that he'd be willing to sleep with the professor if that would help. The frenzy of phoning back and forth across Germany lasted for more than an hour and a half, until Pääbo finally talked to one of the researchers in his own lab. The researcher had actually seen the skullcap and had concluded that it wasn't very old at all. Pääbo immediately lost interest in it.

With old bones, you never really know what you're going to get. A few years ago, Pääbo managed to get hold of a bit of tooth from one of the so-called hobbit skeletons found on the island of Flores, in Indonesia. The hobbits, who were discovered only in 2004, are generally believed to have been diminutive archaic humans—*Homo floresiensis*. The tooth was dated to about seventeen thousand years ago, which meant it was only about half as old as the Croatian Neanderthal bones. But Pääbo couldn't extract any DNA from it.

Then, a year or so later, he obtained a fragment of finger bone that had been unearthed in a cave in southern Siberia along with a weird, vaguely human-looking molar. The finger bone—about the size of a pencil eraser—was more than forty thousand years old. Pääbo assumed that it came either from a modern human or from a Neanderthal. If it proved to be the latter, the site would be the farthest east that Neanderthal remains had been found. In contrast to the hobbit tooth, the finger fragment yielded astonishingly large amounts of DNA. When the analysis of the first bits was completed, Pääbo happened to be in the United States. He called his office, and one of his colleagues said to him, "Are you sitting down?" The DNA showed that the digit did not belong to a modern human *or* to a Neanderthal. Instead, its owner represented an entirely new and previously unsuspected group of hominid. In a paper published in December 2010 in *Nature*, Pääbo dubbed this new group the Denisovans, after the Denisova Cave, where the bone had been found. "Giving Accepted Prehistoric History the Finger," ran one of the newspaper headlines on the discovery. Amazingly—or perhaps, by now, predictably—modern humans must have interbred with Denisovans, too, because contemporary New Guineans carry up to six percent Denisovan DNA.

(Why this is true of New Guineans but not native Siberians or Asians is unclear, but presumably has to do with patterns of human migration.)

With the discovery of the hobbits and the Denisovans, modern humans acquired two new siblings. And it seems likely that as DNA from other old bones is analyzed, additional human relatives will be found; as Chris Stringer, a prominent British paleoanthropologist, put it to me, "I'm sure we've got more surprises to come."

At this point, there's no evidence to indicate what wiped out the Denisovans or the hobbits; however, the timing of their demise and the general pattern of late-Pleistocene extinctions means there's one obvious suspect. Presumably, since they were closely related to us, both Denisovans and hobbits had a long gestation period and therefore shared the megafauna's key vulnerability, a low reproductive rate. All that would have been required to do them in would have been a sustained downward pressure on the number of breeding adults.

And the same holds true for our next-closest kin, which is why, with the exception of humans, all the great apes today are facing oblivion. The number of chimpanzees in the wild has dropped to perhaps half of what it was fifty years ago, and the number of mountain gorillas has followed a similar trajectory. Lowland gorillas have declined even faster; it's estimated the population has shrunk by sixty percent just in the last two decades. Causes of the crash include poaching, disease, and habitat loss; the last of these has been exacerbated by several wars, which have pushed waves of refugees into the gorillas' limited range. Sumatran orangutans are classified as "critically endangered," meaning they're at "extremely high risk of extinction in the wild." In this case, the threat is more peace than violence; most of the remaining orangutans live in the province of Aceh, where a recent end to decades of political unrest has led to a surge in logging, both legal and not. One of the many unintended consequences of the Anthropocene has been the pruning of our own family tree. Having cut down our sister species—the Neanderthals and the Denisovans—many generations ago, we're now working on our first and second cousins. By the time we're done, it's quite possible that there will be among the great apes not a single representative left, except, that is, for us.

ONE of the largest assemblages of Neanderthal bones ever found—remains from seven individuals—was discovered about a century ago at a spot known as La Ferrassie, in southwestern France. La Ferrassie is in the Dordogne, not far from La Chapelle and within half an hour's drive of dozens of other important archaeological sites, including the painted caves at Lascaux. For the last several summers, a team that includes one of Pääbo's colleagues has been excavating at La Ferrassie, and I decided to go down and have a look. I arrived at the dig's headquarters—a converted tobacco barn—just in time for a dinner of boeuf bourguignon, which was served on makeshift tables in the backyard.

The next day, I drove out to La Ferrassie with some of the team's archaeologists. The site lies in a sleepy rural area, right by the side of the road. Many thousands of years ago, La Ferrassie was a huge limestone cave, but one of the walls has since fallen in, and now it is open on two sides. A massive ledge of rock juts out about twenty feet off the ground, like half of a vaulted ceiling. The site is ringed by wire fence and hung with tarps, which give it the aspect of a crime scene.

The day was hot and dusty. Half a dozen students crouched in a long trench, picking at the dirt with trowels. Along the side of the trench, I could see bits of bone sticking out from the reddish soil. The bones toward the bottom, I was told, had been tossed there by Neanderthals. The bones near the top were the leavings of modern humans, who took over the cave once the Neanderthals were gone. The Neanderthal skeletons from the site have long since been removed, but there was still hope that some small bit, like a tooth, might be found. Each bone fragment that was unearthed, along with every flake of flint and anything else that might even remotely be of interest, was set aside to be taken back to the tobacco farm and tagged.

After watching the students chip away for a while, I retreated to the shade. I tried to imagine what life had been like for the Neanderthals at La Ferrassie. Though the area is now wooded, then it would have been treeless. There would have been elk roaming the valley, and reindeer and wild cattle and mammoths. Beyond these stray facts, not much came to me. I put the question to the archaeologists I had driven out with. "It was cold," Shannon McPherron, of the Max Planck Institute, volunteered.

"And smelly," Dennis Sandgathe, of Canada's Simon Fraser University, said.

"Probably hungry," Harold Dibble, of the University of Pennsylvania, added.

"No one would have been very old," Sandgathe said. Later on, back at the barn, I picked through the bits and pieces that had been dug up over the past few days. There were hundreds of fragments of animal bone, each of which had been cleaned and numbered and placed in its own little plastic bag, and hundreds of flakes of flint. Most of the flakes were probably the detritus of toolmaking—the Stone Age equivalent of wood shavings—but some, I learned, were the tools themselves. Once I was shown what to look for, I could see the beveled edges that the Neanderthals had crafted. One tool in particular stood out: a palm-size flint shaped like a teardrop. In archaeological parlance, it was a hand ax, though it probably was not used as an ax in the contemporary sense of the word. It had been found near the bottom of the trench, so it was estimated to be about seventy thousand years old. I took it out of its plastic bag and turned it over. It was almost perfectly symmetrical and—to a human eye, at least—quite beautiful. I said that I thought the Neanderthal who had fashioned it must have had a keen sense of design. McPherron objected.

"We know the end of the story," he told me. "We know what modern culture looks like, and so then what we do is we want to explain how we got here. And there's a tendency to overinterpret the past by projecting the present onto it. So when you see a beautiful hand ax and you say, 'Look at the craftsmanship on this; it's virtually an object of art,' that's your perspective today. But you can't assume what you're trying to prove."

Among the thousands of Neanderthal artifacts that have been unearthed, almost none represent unambiguous attempts at art or adornment, and those that have been interpreted this way—for instance, ivory pendants discovered in a cave in central France—are the subject of endless, often abstruse disputes. (Some archaeologists believe that the pendants were fashioned by Neanderthals who, after coming into contact with modern humans, tried to imitate them. Others argue that the pendants were fashioned by modern humans who occupied the site after the Neanderthals.) This absence has led some to propose that Neanderthals were not capable of art or—what amounts to much the same thing—not interested in it. We may see the hand ax as "beautiful"; they saw it as useful. Genomically speaking, they lacked what might be called the aesthetic mutation.

On my last day in the Dordogne, I went to visit a nearby archaeological site—a human site—called the Grotte des Combarelles. The Grotte is a very narrow cave that zigzags for nearly a thousand feet through a limestone cliff. Since its rediscovery, in the late nineteenth century, the cave has been enlarged and strung with electric lights, which have made it possible to walk through it safely, if not altogether comfortably. When humans first entered the Grotte, twelve or thirteen thousand years ago, it was a different matter. Then the ceiling was so low that the only way to move through the cave would have been to crawl, and the only way to see in the absolute blackness would have been to carry fire. Something—perhaps creativity, perhaps spirituality, perhaps "madness"—drove people along nonetheless. Deep inside the Grotte, the walls are covered with hundreds of engravings. All the images are of animals, many of them now extinct: mammoths, aurochs, woolly rhinos. The most detailed of them possess an uncanny vitality: a wild horse seems to lift its head, a reindeer leans forward, apparently to drink.

It is often speculated that the humans who sketched on the walls of the Grotte des Combarelles thought their images had magical powers, and in a way they were right. The Neanderthals lived in Europe for more than a hundred thousand years and during that period they had no more impact on their surroundings than any other large vertebrate. There is every reason to believe that if humans had not arrived on the scene, the Neanderthals would be there still, along with the wild horses and the woolly rhinos. With the capacity to represent the world in signs and symbols comes the capacity to change it, which, as it happens, is also the capacity to destroy it. A tiny set of genetic variations divides us from the Neanderthals, but that has made all the difference.

CHAPTER XIII

THE THING WITH FEATHERS

Homo sapiens

"FUTUROLOGY HAS NEVER BEEN A VERY RESPECTABLE FIELD OF inquiry," the author Jonathan Schell has written. With this caveat in mind, I've set out for the Institute for Conservation Research, an outpost of the San Diego Zoo thirty miles north of the city. The drive to the institute leads past several golf courses, a winery, and an ostrich farm. When I arrive, the place is hushed, like a hospital. Marlys Houck, a researcher who specializes in tissue culture, leads me down a long corridor into a windowless room. She pulls on a pair of what look like heavy-duty oven mitts and pries the lid off a large metal tank. A ghostly vapor rises from the opening.

At the bottom of the tank is a pool of liquid nitrogen, temperature minus 320 degrees. Suspended above the pool are boxes of little plastic vials. The boxes are stacked in towers, and the vials arranged upright, like pegs, each in its own slot. Houck locates the box she is looking for and counts over several rows, then down. She takes out two of the vials and places them before me on a steel table. "There they are," she says.

Inside the vials is pretty much all that's left of the po`ouli, or black-faced honeycreeper, a chunky bird with a sweet face and a cream-colored chest that lived on Maui. The po`ouli was once described to me as "the most beautiful not particularly beautiful bird in the world," and probably it went extinct a year or two after the San Diego Zoo and the U.S. Fish and Wildlife Service made a last-ditch effort to save it, in the autumn of 2004. At that point, a mere three individuals were known to exist, and the idea was to capture and breed them. But just one bird allowed itself to be netted. It had been thought to be female, but turned out to be male, a development that made Fish and Wildlife Service scientists suspect that only one sex of po`ouli was left. When the captive bird died, the day after Thanksgiving, his body was immediately sent to the San Diego Zoo. Houck raced to the institute to deal with it. "This is our last chance," she remembers thinking. "This is the dodo." Houck succeeded in culturing some of the cells from the bird's eye, and the results of that effort now make up the contents of the vials. She doesn't want the cells to get damaged, so after about a minute she slides the vials back in the box and returns them to the tank.

The windowless room where the po`ouli cells are kept alive—sort of—is called the Frozen Zoo. The name is trademarked, and if other institutions try to use it, they are advised they are breaking the law. The room holds half a dozen tanks just like the one Houck opened, and stored inside of them, in frigid clouds of nitrogen, are cell lines representing nearly a thousand species. (Really this is just half the "zoo"; the other half consists of tanks at a different facility whose location is pointedly kept secret. Each cell line is split between the two facilities, in case the power goes out at one of them.) The Frozen Zoo maintains the world's largest collection of species on ice, but an increasing number of other institutions are also assembling chilled menageries; the Cincinnati Zoo, for example, runs what it calls the CryoBioBank and England's University of Nottingham operates the Frozen Ark.

For now, almost all of the species in deep freeze in San Diego still have flesh-and-blood members. But as more and more plants and animals go the way of the po`ouli, this is likely to change. While Houck is busy resealing the tank, I think of the hundreds of bat corpses collected from the floor of Aeolus Cave that were shipped to the Cryo Collection of the American Museum of Natural History. I try to calculate how many little plastic vials and vats of liquid nitrogen would be required to store cultures of all of the frogs threatened by chytrid and the corals threatened by acidification and the pachyderms threatened by poaching, and the multitudinous species threatened by warming and invasives and fragmentation, and soon I give up; there are too many numbers to keep in my head.

DOES it have to end this way? Does the last best hope for the world's most magnificent creatures—or, for that matter, its least magnificent ones—really lie in pools of liquid nitrogen? Having been alerted to the ways in which we're imperiling other species, can't we take action to protect them? Isn't the whole point of trying to peer into the future so that, seeing dangers ahead, we can change course to avoid them?

Certainly humans can be destructive and shortsighted; they can also be forward-thinking and altruistic. Time and time again, people have demonstrated that they care about what Rachel Carson called "the problem of sharing our earth with other creatures," and that they're willing to make sacrifices on those creatures' behalf. Alfred Newton described the slaughter that was occurring along the British coast; the result was the Act for the Preservation of Sea Birds. John Muir wrote about the damage being done in the mountains of California, and this led to the creation of Yosemite National Park. *Silent Spring* exposed the dangers posed by synthetic pesticides, and within a decade, most uses of DDT had been prohibited. (The fact that there are still bald eagles in the U.S.—indeed the numbers are growing—is one of the many happy consequences of this development.)

Two years after the ban on DDT, Congress in 1974 passed the Endangered Species Act. Since then, the lengths to which people have gone to protect creatures listed under the act is very nearly, in the literal sense of the word, incredible. To cite just one of many possible illustrations, by the mid–nineteen-eighties the population of California condors had dwindled to just twenty-two individuals. To rescue the species—the largest land bird in North America—wildlife biologists raised condor chicks using puppets. They created fake power lines to train the birds not to electrocute themselves; to teach them not to eat trash, they wired garbage to deliver a mild shock. They vaccinated every single condor—today there about four hundred—against West Nile virus, a disease, it's worth noting, for which a human vaccine has yet to be developed. They routinely test the birds for lead poisoning—condors that scavenge deer carcasses often ingest lead shot—and they have treated many of them with chelation therapy. Several condors have been taken in for chelation more than once. The effort to save the whooping crane has involved even more man-hours, most provided by volunteers. Each year, a team of pilots flying ultralight aircraft teaches a new cohort of captive-raised crane chicks how to migrate south for the winter, from Wisconsin to Florida. The journey of nearly thirteen hundred miles can take up to three months, with dozens of stops on private land that owners give over to the birds. Millions of Americans who don't participate directly in such efforts support them indirectly, by joining groups like the World Wildlife Fund, the National Wildlife Federation, Defenders of Wildlife, the Wildlife Conservation Society, the African Wildlife Foundation, the Nature Conservancy, and Conservation International.

Wouldn't it be better, practically and ethically, to focus on what can be done and *is* being done to save species, rather than to speculate gloomily about a future in which the biosphere is reduced to little plastic vials? The director of a conservation group in Alaska once put it to me this way: "People have to have hope. *I* have to have hope. It's what keeps us going."

NEXT door to the Institute for Conservation Research there's a similar looking, dun-colored building that serves as a veterinary hospital. Most of the animals in the hospital, which is also run by the San Diego Zoo, are only passing through, but the building has a permanent resident, too: a Hawaiian crow named Kinohi. Kinohi is one of about a hundred Hawaiian crows, or `alalā, that exist today, all of them in captivity. While in San Diego, I paid a visit to Kinohi with the zoo's director of reproductive physiology, Barbara Durrant, who, I'd been told, was the only person who really understands him. On our way over to see the bird, Durrant stopped off at a commissary of sorts to pick up a selection of his favorite snacks. These included mealworms; a hairless, newborn mouse, known as a "pinky"; and the hindquarters of an adult mouse that had been sliced in half, so that it had a pair of feet on one end and a mess of guts on the other.

No one is sure exactly why the `alalā became extinct in the wild; probably, as with the po'ouli, there are multiple reasons, including habitat loss, predation by invasive species like mongoose, and diseases carried by other invasive species, like mosquitoes. In any event, the last forest-dwelling `alalā is believed to have died in 2002. Kinohi was born at a captive breeding facility on Maui more than twenty years ago. He is, by all accounts, an extremely odd bird. Raised in isolation, he does not identify with other `alalā. Nor does he seem to think of himself as human. "He's in a world all to himself," Durrant told me. "He once fell in love with a spoonbill."

Kinohi was sent to San Diego in 2009 because he refused to mate with any of the other captive crows, and it was decided that something new had to be tried to persuade him to contribute to the species' limited gene pool. It fell to Durrant to figure out how to win Kinohi's heart or, more to the point, his gonads. Kinohi came fairly quickly to accept her attentions—crows do not have phalluses, so Durrant stroked the area around his cloaca—but at the time of my visit he still had failed to deliver what she referred to as "high-quality ejaculate." Another breeding season was approaching, so Durrant was preparing to try again, three times a week for up to five months. If Kinohi ever came through, she was going to rush with his sperm to Maui and try to artificially inseminate one of the females at the breeding facility.

We arrived at Kinohi's cage, which turned out to be more like a suite, with an antechamber large enough for several people to stand in and a back room filled with ropes and other corvid entertainments. Kinohi hopped over to greet us. He was jet black from head to talon. To me, he looked a lot like an average American crow, but Durrant pointed out that he had a much thicker beak and also thicker legs. Kinohi kept his head tilted forward, as if trying

to avoid eye contact. When he saw Durrant, I wondered, did he have the avian equivalent of dirty thoughts? She offered him the snacks she'd brought. He gave a raucous caw that sounded eerily familiar. Crows can mimic human speech, and Durrant translated the caw as "I know."

"I know," Kinohi repeated. "I know."

KINOHI'S tragicomic sex life provides more evidence—if any more was needed—of how seriously humans take extinction. Such is the pain the loss of a single species causes that we're willing to perform ultrasounds on rhinos and handjobs on crows. Certainly the commitment of people like Terri Roth and Barbara Durrant and institutions like the Cincinnati and the San Diego Zoos could be invoked as reason for optimism. And if this were a different kind of book, I would.

Though many of the preceding chapters have been devoted to the extinction (or near-extinction) of individual organisms—the Panamanian golden frog, the great auk, the Sumatran rhino—my real subject has been the pattern they participate in. What I've been trying to do is trace an extinction event—call it the Holocene extinction, or the Anthropocene extinction, or, if you prefer the sound of it, the Sixth Extinction—and to place this event in the broader context of life's history. That history is neither strictly uniformitarian nor catastrophist; rather, it is a hybrid of the two. What this history reveals, in its ups and its downs, is that life is extremely resilient but not infinitely so. There have been very long uneventful stretches and very, very occasionally "revolutions on the surface of the earth."

To the extent that we can identify the causes of these revolutions, they're highly varied: glaciation in the case of the end-Ordovician extinction, global warming and changes in ocean chemistry at the end of the Permian, an asteroid impact in the final seconds of the Cretaceous. The current extinction has its own novel cause: not an asteroid or a massive volcanic eruption but "one weedy species." As Walter Alvarez put it to me, "We're seeing right now that a mass extinction can be caused by human beings."

The one feature these disparate events have in common is change and, to be more specific, rate of change. When the world changes faster than species can adapt, many fall out. This is the case whether the agent drops from the sky in a fiery streak or drives to work in a Honda. To argue that the current extinction event could be averted if people just cared more and were willing to make more sacrifices is not wrong, exactly; still, it misses the point. It doesn't much matter whether people care or don't care. What matters is that people change the world.

This capacity predates modernity, though, of course, modernity is its fullest expression. Indeed, this capacity is probably indistinguishable from the qualities that made us human to begin with: our restlessness, our creativity, our ability to cooperate to solve problems and complete complicated tasks. As soon as humans started using signs and symbols to represent the natural world, they pushed beyond the limits of that world. "In many ways human language is like the genetic code," the British paleontologist Michael Benton has written. "Information is stored and transmitted, with modifications, down the generations. Communication holds societies together and allows humans to escape evolution." Were people simply heedless or selfish or violent, there wouldn't be an Institute for Conservation Research, and there wouldn't be a need for one. If you want to think about why humans are so dangerous to other species, you can picture a poacher in Africa carrying an AK-47 or a logger in the Amazon gripping an ax, or, better still, you can picture yourself, holding a book on your lap.

In the center of the American Museum of Natural History's Hall of Biodiversity, there's an exhibit embedded in the floor. The exhibit is arranged around a central plaque that notes there have been five major extinction events since complex animals evolved, over five hundred million years ago. According to the plaque, "Global climate change and other causes, probably including collisions between earth and extraterrestrial objects," were responsible for these events. It goes on to observe: "Right now we are in the midst of the Sixth Extinction, this time caused solely by humanity's transformation of the ecological landscape."

Radiating out from the plaque are sheets of heavy-duty Plexiglas, and beneath the sheets the fossilized remains of a handful of exemplary casualties. The Plexiglas has been scuffed by the shoes of the tens of thousands of museum visitors who have walked across it, probably for the most part oblivious of what's beneath their feet. But crouch down and look closely and you can see that each of the fossils is labeled with the name of the species as well as the extinction event that brought its lineage to an end. The fossils are arranged in chronological order, so that the oldest—graptolites from the Ordovician—are close to the center, while the youngest—*Tyrannosaurus rex* teeth from the late Cretaceous—are farther away. If you stand at the edge of the exhibit, which is really the only place from which to view it, you are positioned right where the victims of the Sixth Extinction should go.

In an extinction event of our own making, what happens to us? One possibility—the possibility implied by the Hall of Biodiversity—is that we, too, will eventually be undone by our "transformation of the ecological landscape." The logic behind this way of thinking runs as follows: having freed ourselves from the constraints of evolution, humans nevertheless remain dependent on the earth's biological and geochemical systems. By disrupting these systems—cutting down tropical rainforests, altering the composition of the atmosphere, acidifying the oceans—we're putting our own survival in danger. Among the many lessons that emerge from the geologic record, perhaps the most sobering is that in life, as in mutual funds, past performance is no guarantee of future results. When a mass extinction occurs, it takes out the weak and also lays low the strong. V-shaped graptolites were everywhere, and then they were nowhere. Ammonites swam around for hundreds of millions of years, and then they were gone. The anthropologist Richard Leakey has warned that "*Homo sapiens* might not only be the agent of the sixth extinction, but also risks being one of its victims." A sign in the Hall of Biodiversity offers a quote from the Stanford ecologist Paul Ehrlich: IN PUSHING OTHER SPECIES TO EXTINCTION, HUMANITY IS BUSY SAWING OFF THE LIMB ON WHICH IT PERCHES.

Another possibility—considered by some to be more upbeat—is that human ingenuity will outrun any disaster human ingenuity sets in motion. There are serious scientists who argue, for instance, that should global warming become too grave a threat, we can counteract it by reengineering the atmosphere. Some schemes involve scattering sulfates into the stratosphere to reflect sunlight back out to space; others involve shooting water droplets over the Pacific to brighten clouds. If none of this works and things really go south, there are those who maintain people will still be OK; we'll simply decamp to other planets. One recent book advises building cities "on Mars, Titan, Europa, the moon, asteroids, and any other uninhabited chunk of matter we can find."

"Don't worry," its author observes. "As long as we keep exploring, humanity is going to survive."

Obviously, the fate of our own species concerns us disproportionately. But at the risk of sounding anti-human—some of my best friends are humans!—I will say that it is not, in the end, what's most worth attending to. Right now, in the amazing moment that to us counts as the present, we are deciding, without quite meaning to, which evolutionary pathways will remain open and which will forever be closed. No other creature has ever managed this, and it will, unfortunately, be our most enduring legacy. The Sixth Extinction will continue to determine the course of life long after everything people have written and painted and built has been ground into dust and giant rats have—or have not—inherited the earth.

SOME MAJOR EVENTS IN THE HISTORY OF LIFE—THE LAST HALF-BILLION YEARS

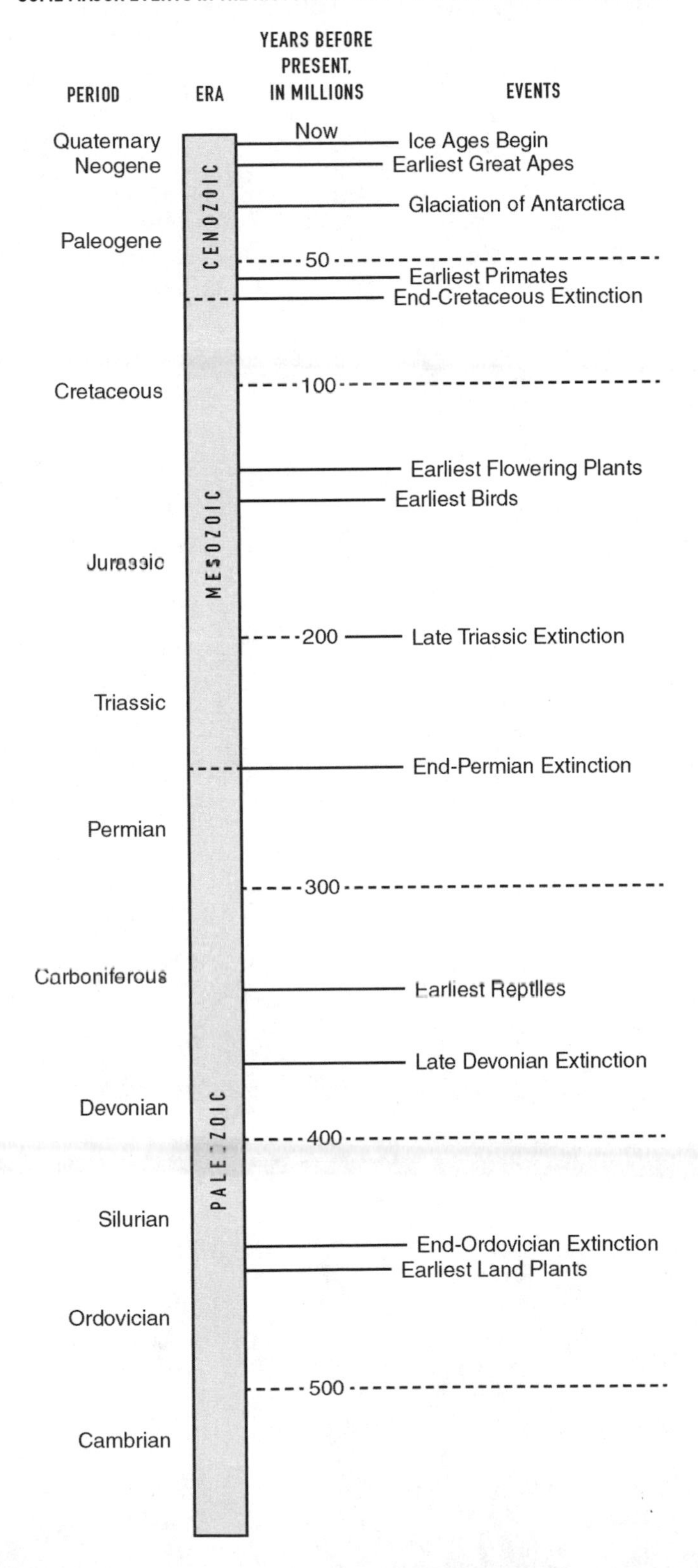

ACKNOWLEDGMENTS

A journalist writing a book about mass extinction needs a lot of help. Many very knowledgeable, generous, and patient people lent their time and expertise to this project.

For help understanding what's come to be known as the amphibian crisis, I am indebted to Edgardo Griffith, Heidi Ross, Paul Crump, Vance Vredenburg, David Wake, Karen Lips, Joe Mendelson, Erica Bree Rosenblum, and Allan Pessier

For his behind-the-scenes tour of Paris's Museum of Natural History, I want to thank Pascal Tassy. For showing me the great auk and its former haunts, I'd like to thank Guðmundur Guðmundsson, Reynir Sveinsson, and Halldór Ármannsson, as well as Magnus Bernhardsson, who made the trip out to Eldey possible. Neil Landman generously showed me the Cretaceous sites of New Jersey and his extraordinary collection of ammonites. Thanks to Lindy Elkins-Tanton and Andy Knoll for sharing their expertise on the end-Permian extinction, and to Nick Longrich and Steve D'Hondt for sharing theirs on the end-Cretaceous.

I owe a special debt to Jan Zalasiewicz, who, in addition to taking me graptolite hunting in Scotland, has answered innumerable questions over the last few years. I am also grateful to Dan Condon and Ian Millar for a memorable (if wet) expedition, and to Paul Crutzen for explaining to me his idea of the Anthropocene.

Ocean acidification is a daunting topic. I never would have been able to write about it without the help of Chris Langdon, Richard Feely, Chris Sabine, Joanie Kleypas, Victoria Fabry, Ulf Riebesell, Lee Kump, and Mark Pagani. I am especially grateful to Jason Hall-Spencer, who took me swimming out to Castello Aragonese in the freezing cold and afterward patiently answered my many questions. Also many thanks to Maria Cristina Buia for arranging the journey.

I have turned to Ken Caldeira over and over again for help understanding topics in climate science and marine chemistry. I am deeply indebted to him and to his wife, Lilian, and to the whole team I met at One Tree: Jack Silverman, Kenny Schneider, Tanya Rivlin, Jen Reiffel, and the inimitable Russell Graham. Thanks also to Davey Kline, Brad Opdyke, Selina Ward, and Ove Hoegh-Guldberg.

Miles Silman was an extraordinary guide to an extraordinary part of the world. I can't thank him enough for sharing so much of his time and his knowledge. I'd also like to express my gratitude to his doctoral students, William Farfan Rios and Karina Garcia Cabrera. Many thanks, too, to Chris Thomas.

This book might never even have been attempted without the help of Tom Lovejoy. His generosity and patience are, as far as I can tell, boundless, and I am deeply grateful to him for his assistance and his encouragement. Mario Cohn-Haft was an expert and wonderfully good-humored guide to the Amazon rainforest. I also want to thank Rita Mesquita, José Luís Camargo, Gustavo Fonseca, and Virgilio Viana.

Scott Darling and Al Hicks were among the very first people to appreciate the seriousness of white-nose syndrome. They shared with me what they were learning as they were learning it and helped me enormously. Ryan Smith, Susi von Oettingen, and Alyssa Bennett were kind enough to keep taking me back up to Aeolus. Joe Roman generously read and commented on the section of the book on invasive species.

Terri Roth and Chris Johnson helped me to understand megafauna, past and present. A special thanks to John Alroy for his calculations on extinction rates, and thanks, too, to Anthony Barnosky.

Svante Pääbo spent many hours explaining to me the complexities of paleogenetics in general and of the Neanderthal Genome Project in particular. I want to thank him as well as Shannon McPherron, who graciously showed me around La Ferrassie, and Ed Green, who was always willing to answer one last question.

Marlys Houck, Oliver Ryder, Barbara Durrant, and Jenny Mehlow were very generous to me when I visited San Diego.

I'd like to thank the reference librarians at Williams College, who tracked down books and articles that were practically untrackable, and Jay Pasachoff, who kindly lent me his files on the end-Cretaceous extinction.

In 2010, I was fortunate enough to receive a fellowship from the John Simon Guggenheim Memorial Foundation, which allowed me to travel to places I would otherwise not have been able to go. Indirect support for this project came from a Lannan Literary Fellowship and from the Heinz Family Foundation.

Parts of several chapters of this book first appeared in the *New Yorker*. For their counsel and support and for their patience, I am deeply indebted to David Remnick and Dorothy Wickenden. For his always sage advice, I want to thank John Bennet. Parts of other chapters ran in *National Geographic* and on the Web site e360. I want to thank Rob Kunzig, Jamie Shreeve, and Roger Cohn for their help and their ideas. Also many thanks to Steven Barclay and Eliza Fischer for their unflagging support.

Thanks to Laura Wyss, Meryl Levavi, Caroline Zancan, and Vicki Haire for transforming an unruly manuscript into a book.

Gillian Blake was the best sort of editor one could hope for on a project like this: smart, probing, and unflappable. Whenever things seemed to be veering off the rails, she calmly guided them back. Kathy Robbins was, as always, matchless. Her counsel and insight were invaluable, and her good cheer was unfailing.

Many friends and family members helped with this multiyear endeavor, some of them probably without even realizing it. Thanks to Jim and Karen Shepard, Andrea Barrett, Susan Greenfield, Todd Purdum, Nancy Pick, Lawrence Douglas, and Stewart Adelson, and to Marlene, Gerald, and Dan Kolbert. A special thanks to Barry Goldstein. Thanks, too, to Ned Kleiner, who assisted in pulling together the very last pieces of this book, and to Aaron and Matthew Kleiner, who never guilt-tripped their mother for not attending their soccer games.

Finally, I want to thank my husband, John Kleiner, who once again helped in more ways than he should have. I wrote this book with him and for him.

NOTES

CHAPTER 1: THE SIXTH EXTINCTION

5 **I first read about the frogs:** Ruth A. Musgrave, "Incredible Frog Hotel," *National Geographic Kids,* Sept. 2008, 16–19.

6 **I ran across another frog-related article:** D. B. Wake and V. T. Vredenburg, "Colloquium Paper: Are We in the Midst of the Sixth Mass Extinction? A View from the World of Amphibians," *Proceedings of the National Academy of Sciences* 105 (2008): 11466–73.

12 **In the late nineteen-eighties, an American herpetologist:** Martha L. Crump, *In Search of the Golden Frog* (Chicago: University of Chicago Press, 2000), 165.

15 **For what's probably the best-studied group:** I am indebted to John Alroy for walking me through the complexities of calculating background extinction rates. See also Alroy's "Speciation and Extinction in the Fossil Record of North American Mammals," in *Speciation and Patterns of Diversity,* edited by Roger Butlin, Jon Bridle, and Dolph Schluter (Cambridge: Cambridge University Press, 2009), 310–23.

16 **Anthony Hallam and Paul Wignall, British paleontologists:** A. Hallam and and P. B. Wignall, *Mass Extinctions and Their Aftermath* (Oxford: Oxford University Press, 1997), 1.

16 **Another expert, David Jablonski:** David Jablonski, "Extinctions in the Fossil Record," in *Extinction Rates,* edited by John H. Lawton and Robert M. May (Oxford: Oxford University Press, 1995), 26.

16 **Michael Benton, a paleontologist:** Michael Benton, *When Life Nearly Died:*

The Greatest Mass Extinction of All Time (New York: Thames and Hudson, 2003), 10.

16 **A fifth paleontologist, David Raup:** David M. Raup, *Extinction: Bad Genes or Bad Luck?* (New York: Norton, 1991), 84.

17 **Almost certainly, though, the rate is lower:** John Alroy, personal communication, June 9, 2013.

17 **"I sought a career":** Joseph R. Mendelson, "Shifted Baselines, Forensic Taxonomy, and Rabb's Fringe-limbed Treefrog: The Changing Role of Biologists in an Era of Amphibian Declines and Extinctions," *Herpetological Review* 42 (2011): 21–25.

17 **it's been calculated that the group's extinction rate:** Malcolm L. McCallum, "Amphibian Decline or Extinction? Current Declines Dwarf Background Extinction Rates," *Journal of Herpetology* 41 (2007): 483–91.

17 **It is estimated that one-third of reef-building corals:** Michael Hoffmann et al., "The Impact of Conservation on the Status of the World's Vertebrates," *Science* 330 (2010): 1503–9. See also *Spineless—Status and Trends of the World's Invertebrates,* a report from the Zoological Society of London, published Aug. 31, 2012.

CHAPTER II: THE MASTODON'S MOLARS

25 **it was labeled the "tooth of a Giant":** Paul Semonin, *American Monster: How the Nation's First Prehistoric Creature Became a Symbol of National Identity* (New York: New York University Press, 2000), 15.

25 **On one leg, a French soldier:** Frank H. Severance, *An Old frontier of France: The Niagara Region and Adjacent Lakes under French Control* (New York: Dodd, 1917), 320.

26 **"What animal does it come from?":** Quoted in Claudine Cohen, *The Fate of the Mammoth: Fossils, Myth, and History* (Chicago: University of Chicago Press, 2002), 90.

27 **"The supposed American elephant":** Quoted in Semonin, *American Monster,* 147–48.

27 **With great trepidation, Buffon allowed:** Cohen, *The Fate of the Mammoth,* 98.

28 **a temperament one friend compared:** Quoted in Dorinda Outram, *Georges Cuvier: Vocation, Science and Authority in Post-Revolutionary France* (Manchester, England: Manchester University Press, 1984), 13.

28 **An older colleague would later describe him:** Quoted in Martin J. S. Rudwick, *Bursting the Limits of Time: The Reconstruction of Geohistory in the Age of Revolution* (Chicago: University of Chicago Press, 2005), 355.

28 **On the basis of his examination:** Rudwick, *Bursting the Limits of Time,* 361.

29 **"It is to anatomy alone":** Georges Cuvier and Martin J. S. Rudwick, *Georges Cuvier, Fossil Bones, and Geological Catastrophes: New Translations and Interpretations of the Primary Texts* (Chicago, University of Chicago Press, 1997), 19.

33 **During the Revolution, Cuvier was thin:** Quoted in Stephen Jay Gould, *The Panda's Thumb: More Reflections in Natural History* (New York: Norton, 1980), 146.

34 **"I should say that I have been supported":** Cuvier and Rudwick, *Fossil Bones*, 49.

34 **"If so many lost species":** Ibid., 56.

35 **When he uncovered the fossil animal's forelimbs:** Rudwick, *Bursting the Limits of Time*, 501.

36 **To publicize the exhibition:** Charles Coleman Sellers, *Mr. Peale's Museum: Charles Willson Peale and the First Popular Museum of Natural Science and Art* (New York: Norton, 1980), 142.

37 **the newspapers reported on:** Charles Willson Peale, *The Selected Papers of Charles Willson Peale and His Family*, edited by Lillian B. Miller, Sidney Hart, and David C. Ward, vol. 2, pt. 1 (New Haven, Conn.: Yale University Press, 1988), 408.

38 **He wrote to Jefferson:** Ibid., vol. 2, pt. 2, 1189.

38 **Jefferson was lukewarm:** Ibid., vol. 2, pt. 2, 1201.

38 **"Is not Cuvier":** Quoted in Toby A. Appel, *The Cuvier-Geoffroy Debate: French Biology in the Decades before Darwin* (New York: Oxford University Press, 1987), 190.

40 **"One shouldn't anticipate":** Quoted in Martin J. S. Rudwick, *Worlds Before Adam: The Reconstruction of Geohistory in the Age of Reform* (Chicago: University of Chicago Press, 2008), 32.

41 **"be constructed for devouring prey":** Cuvier and Rudwick, *Fossil Bones*, 217.

42 **"ducks by dint of diving":** Quoted in Richard Wellington Burkhardt, *The Spirit of System: Lamarck and Evolutionary Biology* (Cambridge, MA: Harvard University Press, 1977), 199.

43 **Among the crates of loot:** Cuvier and Rudwick, *Fossil Bones*, 229.

43 **Lamarck objected:** Rudwick, *Bursting the Limits of Time*, 398.

43 **"I know that some naturalists":** Cuvier and Rudwick, *Fossil Bones*, 228.

43 **Like the "enchanted palaces":** Georges Cuvier, "Elegy of Lamarck," *Edinburgh New Philosophical Journal* 20 (1836): 1–22.

44 **"Living organisms without number":** Cuvier and Rudwick, *Fossil Bones*, 190.

44 **organisms simply moved on:** Ibid., 261.

CHAPTER III: THE ORIGINAL PENGUIN

47 **When he proposed it:** Rudwick, *Worlds Before Adam*, 358.

47 **Lyell had grown up:** Leonard G. Wilson, "Lyell: The Man and His Times," in *Lyell: The Past Is the Key to the Present*, edited by Derek J. Blundell and Andrew C. Scott (Bath, England: Geological Society, 1998), 21.

48 **"very obliging":** Charles Lyell, *Life, Letters and Journals of Sir Charles Lyell*, edited by Mrs. Lyell, vol. 1 (London: John Murray, 1881), 249.

48 **"the huge iguanodon":** Charles Lyell. *Principles of Geology*, vol. 1 (Chicago: University of Chicago Press, 1990), 123.

48 **"that there is no foundation":** Ibid., vol. 1, 153.

50 **Lyell boasted:** Leonard G. Wilson, *Charles Lyell, the Years to 1841: The Revolution in Geology* (New Haven, Conn.: Yale University Press, 1972), 344.

50 **when he spoke in Boston:** A. Hallam, *Great Geological Controversies* (Oxford: Oxford University Press, 1983), ix.

50 **He produced a cartoon:** For a discussion of the meaning of the cartoon, see Martin J. S. Rudwick, *Lyell and Darwin, Geologists: Studies in the Earth Sciences in the Age of Reform* (Aldershot, England: Ashgate, 2005), 537–40.

51 **In fact, Darwin developed his theory:** Frank J. Sulloway, "Darwin and His Finches: The Evolution of a Legend," *Journal of the History of Biology* 15 (1982): 1–53.

51 **Lyell further contended:** Lyell, *Principles of Geology*, vol. 1, 476.

51 **caught up with him in the Falklands:** Sandra Herbert, *Charles Darwin, Geologist* (Ithaca, N.Y.: Cornell University Press, 2005), 63.

52 **Recent searches:** Claudio Soto-Azat et al., "The Population Decline and Extinction of Darwin's Frogs," *PLOS ONE* 8 (2013).

53 **Darwin presented them to Lyell:** David Dobbs, *Reef Madness: Charles Darwin, Alexander Agassiz, and the Meaning of Coral* (New York: Pantheon, 2005), 152.

53 **Lyell "recognized that Darwin":** Rudwick, *Worlds before Adam*, 491.

53 **"Without Lyell":** Janet Browne, *Charles Darwin: Voyaging* (New York: Knopf, 1995), 186.

53 **"a single pair":** Charles Lyell, *Principles of Geology*, vol. 2 (Chicago: University of Chicago Press, 1990), 124.

53 **"exactly the kind of miracle":** Ernst Mayr, *The Growth of Biological Thought: Diversity, Evolution, and Inheritance* (Cambridge, Mass.: Belknap Press of Harvard University Press, 1982), 407.

54 **"It may be said that":** Charles Darwin, *On the Origin of Species: A Facsimile of the First Edition* (Cambridge, Mass.: Harvard University Press, 1964), 84.

54 **"The appearance of new forms":** Ibid., 320.

54 **"The theory of natural selection":** Ibid., 320.

55 **"The complete extinction of the species":** Ibid., 318.

56 **Icelandair provided:** Errol Fuller, *The Great Auk* (New York: Abrams, 1999), 197.

58 **genetic analysis has shown:** Truls Moum et al., "Mitochondrial DNA Sequence Evolution and Phylogeny of the Atlantic Alcidae, Including the Extinct Great Auk (Pinguinus impennis)," *Molecular Biology and Evolution* 19 (2002): 1434–39.

58 **the birds' "astonishing velocity":** Jeremy Gaskell, *Who Killed the Great Auk?* (Oxford: Oxford University Press, 2000), p. 8.

59 **great auk bones have been found:** Ibid., 9.

59 **"They are always in the water":** Quoted in Fuller, *The Great Auk*, 64.

60 **"by hundreds at a time":** Quoted in Gaskell, *Who Killed the Great Auk?*, 87.

60 **"the great auks of Funk Island":** Fuller, *The Great Auk*, 64.

60 **"You take a kettle":** Quoted in ibid., 65–66.

60 **It's been estimated:** Tim Birkhead, "How Collectors Killed the Great Auk," *New Scientist* 142 (1994): 26.

61 **"The destruction which they have made":** Quoted in Gaskell, *Who Killed the Great Auk?*, 109.

62 **"rare and accidental":** Quoted in ibid., 37. Gaskell also points out the contradiction in Audubon's description.

66 **Subsequent detective work:** Fuller, *The Great Auk*, 228–29.

66 **"It was with heavy hearts":** Alfred Newton, "Abstract of Mr. J. Wolley's Researches in Iceland Respecting the Gare-Fowl or Great Auk," *Ibis* 3 (1861): 394.

66 **"peculiar attraction":** Alexander F. R. Wollaston, *Life of Alfred Newton* (New York: E. P. Dutton, 1921), 52.

67 **"It came to me like the direct revelation":** Quoted in ibid., 112.

67 **"pure and unmitigated Darwinism":** Quoted in ibid., 121.

67 **It is not mentioned:** Many, but not all, of Darwin's letters are available to the public online; Elizabeth Smith of the Darwin Correspondence Project kindly performed a search of the entire database.

68 **On this basis, Lawson claimed:** Thalia K. Grant and Gregory B. Estes, *Darwin in Galápagos: Footsteps to a New World* (Princeton, N.J.: Princeton University Press, 2009), 123.

68 **In fact, it probably disappeared:** Ibid., 122.

69 **"the denial of humanity's special status":** David Quammen, *The Reluctant Mr. Darwin: An Intimate Portrait of Charles Darwin and the Making of His Theory of Evolution* (New York: Atlas Books/Norton, 2006), 209.

CHAPTER IV: THE LUCK OF THE AMMONITES

72 **"In science, sometimes it's better":** Walter Alvarez, "Earth History in the Broadest Possible Context," Ninety-Seventh Annual Faculty Research Lecture, University of California, Berkeley, International House, delivered Apr. 29, 2010.

74 **"hard-core uniformitarianism":** Walter Alvarez, *T. rex and the Crater of Doom* (Princeton, N.J.: Princeton University Press, 1997), 139.

75 **The amount of iridium:** Ibid., 69.

75 **"like a shark smelling blood":** Richard Muller, *Nemesis* (New York: Weidenfeld and Nicolson, 1988), 51.

76 **"to connect the dinosaurs":** Quoted in Charles Officer and Jake Page, "The K-T Extinction," in *Language of the Earth: A Literary Anthology*, 2nd ed., edited by Frank H. T. Rhodes, Richard O. Stone, and Bruce D. Malamud (Chichester, England: Wiley, 2009), 183.

77 **"The Cretaceous extinctions were gradual":** Quoted in Malcolm W. Browne, "Dinosaur Experts Resist Meteor Extinction Idea," *New York Times*, Oct. 29, 1985.

77 **"Astronomers should leave":** *New York Times* Editorial Board, "Miscasting the Dinosaur's Horoscope," *New York Times*, Apr. 2, 1985.

78 **he noted a "chasm":** Lyell, *Principles of Geology*, vol. 3 (Chicago: University of Chicago Press, 1991), 328.

78 **Rudists have been described:** David M. Raup, *The Nemesis Affair: A Story of the Death of Dinosaurs and the Ways of Science* (New York: Norton, 1986), 58.

79 **"I look at the natural geological record":** Darwin, *On the Origin of Species*, 310–11.

79 **"So profound is our ignorance":** Ibid., 73.

79 **"a long and essentially continuous process":** George Gaylord Simpson, *Why and How: Some Problems and Methods in Historical Biology* (Oxford: Pergamon Press, 1980), 35.

80 **"codswallop":** Quoted in Browne, "Dinosaur Experts Resist Meteor Extinction Idea."

80 **In 1984, grains of shocked quartz:** B. F. Bohor et al., "Mineralogic Evidence for an Impact Event at the Cretaceous-Tertiary Boundary," *Science* 224 (1984): 867–69.

85 **In a drawing that accompanies:** Neil Landman et al., "Mode of Life and Habitat of Scaphitid Ammonites," *Geobios* 54 (2012): 87–98.

86 **"Basically, if you were a triceratops":** Personal communication, Steve D'Hondt, Jan. 5, 2012.

87 **Birds were also hard-hit:** Nicholas R. Longrich, T. Tokaryk, and D. J. Field, "Mass Extinction of Birds at the Cretaceous–Paleogene (K–Pg) Boundary," *Proceedings of the National Academy of Sciences* 108 (2011): 15253–257.

87 **The same goes for lizards:** Nicholas R. Longrich, Bhart-Anjan S. Bhullar, and Jacques A. Gauthier, "Mass Extinction of Lizards and Snakes at the Cretaceous-Paleogene Boundary," *Proceedings of the National Academy of Sciences* 109 (2012): 21396–401.

87 **Mammals' ranks, too, were devastated:** Kenneth Rose, *The Beginning of the Age of Mammals* (Baltimore: Johns Hopkins University Press, 2006), 2.

90 **"the rules of the survival game":** Paul D. Taylor, *Extinctions in the History of Life* (Cambridge: Cambridge University Press, 2004), 2.

CHAPTER V: WELCOME TO THE ANTHROPOCENE

92 **Others were completely flummoxed:** Jerome S. Bruner and Leo Postman, "On the Perception of Incongruity: A Paradigm," *Journal of Personality* 18 (1949): 206–23. I am indebted to James Gleick for drawing my attention to this experiment: see *Chaos: Making a New Science* (New York: Viking, 1987), 35.

93 **"In science, as in the playing card experiment":** Thomas S. Kuhn, *The Structure of Scientific Revolutions*, 2nd ed. (Chicago: University of Chicago Press, 1970), 64.

94 **"God knows how many catastrophes":** Quoted in Patrick John Boylan, "William Buckland, 1784–1859: Scientific Institutions, Vertebrate Paleontology and Quaternary Geology" (Ph.D. dissertation, University of Leicester, England, 1984), 468.

94 **"as explosive for science":** William Glen, *Mass Extinction Debates: How Science Works in a Crisis* (Stanford, Calif.: Stanford University Press, 1994), 2.

96 **Something like eighty-five percent:** Hallam and Wignall, *Mass Exinctions and Their Aftermath*, 4.

97 **"Had the list of survivors":** Richard A. Fortey, *Life: A Natural History of the First Four Billion Years of Life on Earth* (New York: Vintage, 1999), 135.

101 **The two paleontologists:** David M. Raup and J. John Sepkoski Jr., "Periodicity of Extinctions in the Geologic Past," *Proceedings of the National Academy of Sciences* 81 (1984): 801–5.

102 **everything but sex and the royal family:** Raup, *The Nemesis Affair*, 19.

102 **another disapproving editorial:** *New York Times* Editorial Board, "Nemesis of Nemesis," *New York Times*, July 7, 1985.

102 **"We felt sure that there would be":** Luis W. Alvarez, "Experimental Evidence That an Asteroid Impact Led to the Extinction of Many Species 65 Million Years Ago," *Proceedings of the National Academy of Sciences* 80 (1983): 633.

103 **One theory has it:** Timothy M. Lenton et al., "First Plants Cooled the Ordovician," *Nature Geoscience* 5 (2012): 86–89.

103 **the seas warmed:** Timothy Kearsey et al., "Isotope Excursions and Palaeotemperature Estimates from the Permian/Triassic Boundary in the Southern Alps (Italy)," *Palaeogeography, Palaeoclimatology, Palaeoecology* 279 (2009): 29–40.

103 **the whole episode lasted:** Shu-zhong Shen et al., "Calibrating the End-Permian Mass Extinction," *Science* 334 (2011): 1367–72.

104 **One hypothesis has it:** Lee R. Kump, Alexander Pavlov, and Michael A. Arthur, "Massive Release of Hydrogen Sulfide to the Surface Ocean and Atmosphere during Intervals of Oceanic Anoxia," *Geology* 33 (2005): 397–400.

104 **"truly grotesque place":** Carl Zimmer, introduction to paperback edition of *T. Rex and the Crater of Doom* (Princeton, N.J.: Princeton University Press, 2008), xv.

105 **not much thicker than a cigarette paper:** Jan Zalasiewicz, *The Earth After Us: What Legacy Will Humans Leave in the Rocks?* (Oxford: Oxford University Press, 2008), 89.

105 **"We have already left a record":** Ibid., 240.

106 **"a grey tide":** Quoted in William Stolzenburg, *Rat Island: Predators in Paradise and the World's Greatest Wildlife Rescue* (New York: Bloomsbury, 2011), 21.

106 **A recent study of pollen:** Terry L. Hunt, "Rethinking Easter Island's Ecological Catastrophe," *Journal of Archaeological Science* 34 (2007): 485–502.

106 **"a species or two of large naked rodent":** Zalasiewicz, *The Earth After Us*, 9.

108 **"Because of these anthropogenic emissions":** Paul J. Crutzen, "Geology of Mankind," *Nature* 415 (2002): 23.

109 **"as future evolution":** Jan Zalasiewicz et al., "Are We Now Living in the Anthropocene?" *GSA Today* 18 (2008): 6.

CHAPTER VI: THE SEA AROUND US

117 **the tally they came up with was very different:** Jason M. Hall-Spencer et al., "Volcanic Carbon Dioxide Vents Show Ecosystem Effects of Ocean Acidification," *Nature* 454 (2008): 96–99. Details from supplementary tables.

119 **in one mesocosm experiment:** Ulf Reibesell, personal communication, Aug. 6, 2012.

120 **There's strong evidence:** Wolfgang Kiessling and Carl Simpson, "On the Potential for Ocean Acidification to Be a General Cause of Ancient Reef Crises," *Global Change Biology* 17 (2011): 56–67.

121 **It's been estimated that calcification evolved:** Andrew H. Knoll, "Biomineralization and Evolutionary History," *Reviews in Mineralogy and Geochemistry* 54 (2003): 329–56.

122 **three-quarters of the missing:** Hall-Spencer et al., "Volcanic Carbon Dioxide Vents Show Ecosystem Effects of Ocean Acidification," *Nature* 454 (2008): 96–99.

123 **This comes to a stunning:** For up-to-date figures on atmospheric emissions and ocean uptake of carbon dioxide, thanks to Chris Sabine of NOAA's PMEL Carbon Program.

123 **"Time is the essential ingredient":** Rachel Carson, *Silent Spring*, 40th anniversary ed. (Boston: Houghton Mifflin, 2002), 6.

123 **But even this spectacular event:** Jennifer Chu, "Timeline of a Mass Extinction," MIT News Office, published online Nov. 18, 2011.

124 **"It is the rate":** Lee Kump, Timothy Bralower, and Andy Ridgwell, "Ocean Acidification in Deep Time," *Oceanography* 22 (2009): 105.

CHAPTER VII: DROPPING ACID

128 **"a wall of Coral Rock":** Quoted in James Bowen and Margarita Bowen, *The Great Barrier Reef: History, Science, Heritage* (Cambridge: Cambridge University Press, 2002), 11.

128 **"thrown up to such a height?":** Quoted in ibid., 2.

128 **Lyell's theory:** Dobbs, *Reef Madness*, 147–48. Lyell mistakenly attributed the idea to Adelbert von Chamisso, a naturalist who accompanied Otto von Kotzebue.

129 **"astoundingly correct":** Ibid., 256.

130 **"It is likely that reefs":** Charles Sheppard, Simon K. Davy, and Graham M. Pilling, *The Biology of Coral Reefs* (Oxford: Oxford University Press, 2009), 278.

130 **"rapidly eroding rubble banks":** Ove Hoegh-Guldberg et al., "Coral Reefs under Rapid Climate Change and Ocean Acidification," *Science* 318 (2007): 1737–42.

132 **Caldeira published the first part of his paper:** Ken Caldeira and Michael E. Wickett, "Anthropogenic Carbon and Ocean pH," *Nature* 425 (2003): 365.

137 **The experiment, modeled on Hall-Spencer's work:** Katherina E. Fabricius et al., "Losers and Winners in Coral Reefs Acclimatized to Elevated Carbon Dioxide Concentrations," *Nature Climate Change* 1 (2011): 165–69.

138 **"A few decades ago":** J. E. N. Veron, "Is the End in Sight for the World's Coral Reefs?" *e360*, published online Dec. 6, 2010.

138 **A recent study by a team of Australian researchers:** Glenn De'ath et al., "The 27-Year Decline of Coral Cover on the Great Barrier Reef and Its Causes," *Proceedings of the National Academy of Sciences* 109 (2012): 17995–99.

138 **"all coral reefs will cease to grow":** Jacob Silverman et al., "Coral Reefs May Start Dissolving when Atmospheric CO_2 Doubles," *Geophysical Research Letters* 35 (2009).

139 **in a square meter's worth:** Laetitia Plaisance et al., "The Diversity of Coral Reefs: What Are We Missing?" *PLOS ONE* 6 (2011).

142 **"that of most terrestrial animal groups":** Kent E. Carpenter et al., "One-Third of Reef-Building Corals Face Elevated Extinction Risk from Climate Change and Local Impacts," *Science* 321 (2008): 560–63.

143 **"An island slumbering":** By June Chilvers, reprinted in Harold Heatwole, Terence Done, and Elizabeth Cameron, *Community Ecology of a Coral Cay: A Study of One-Tree Island, Great Barrier Reef, Australia* (The Hague: W. Junk, 1981), v.

CHAPTER VIII: THE FOREST AND THE TREES

151 **realize "that you are standing":** Barry Lopez, *Arctic Dreams* (1986; reprint, New York: Vintage, 2001), 29.

152 **Massachusetts around fifty-five:** Gordon P. DeWolf, *Native and Naturalized Trees of Massachusetts* (Amherst: Cooperative Extension Service, University of Massachusetts, 1978).

152 **though not, interestingly enough, for aphids:** John Whitfield, *In the Beat of a Heart: Life, Energy, and the Unity of Nature* (Washington, D.C.: National Academies Press, 2006), 212.

152 **"a spectacle as varied":** Alexander von Humboldt and Aimé Bonpland, *Essay on the Geography of Plants,* edited by Stephen T. Jackson, translated by Sylvie Romanowski (Chicago: University of Chicago Press, 2008), 75.

152 **"The verdant carpet":** Alexander von Humboldt, *Views of Nature, or, Contemplations on the Sublime Phenomena of Creation with Scientific Illustrations,* translated by Elsie C. Otté and Henry George Bohn (London: H. G. Bohn, 1850), 213–17.

153 **One theory holds:** Many theories of the latitudinal diversity gradient are summarized in Gary G. Mittelbach et al., "Evolution and the Latitudinal Diversity Gradient: Speciation, Extinction and Biogeography," *Ecology Letters* 10 (2007): 315–31.

153 **A famous paper:** Daniel H. Janzen, "Why Mountain Passes Are Higher in the Tropics," *American Naturalist* 101 (1967): 233–49.

153 **"evolution has had a fair chance":** Alfred R. Wallace, *Tropical Nature and Other Essays* (London: Macmillan, 1878), 123.

159 **Trees in *Schefflera*:** Kenneth J. Feeley et al., "Upslope Migration of Andean Trees," *Journal of Biogeography* 38 (2011): 783–91.

160 **"some of the most acute and powerful intellects":** Alfred R. Wallace, *The Wonderful Century: Its Successes and Its Failures* (New York: Dodd, Mead, 1898), 130.

161 **"As the cold came on":** Darwin, *On the Origin of Species*, 366–67.

162 **the Andes are expected:** Rocío Urrutia and Mathias Vuille, "Climate Change Projections for the Tropical Andes Using a Regional Climate Model: Temperature and Precipitation Simulations for the End of the 21st Century," *Journal of Geophysical Research* 114 (2009).

162 **I'd recently read a paper:** Alessandro Catenazzi et al., "*Batrachochytrium dendrobatidis* and the Collapse of Anuran Species Richness and Abundance in the Upper Manú National Park, Southeastern Peru," *Conservation Biology* 25 (2011): 382–91.

167 **"Here's another way to express":** Anthony D. Barnosky, *Heatstroke: Nature in an Age of Global Warming* (Washington, D.C.: Island Press/Shearwater Books, 2009), 55–56.

167 **The study ran as the cover article:** Chris D. Thomas et al., "Extinction Risk from Climate Change," *Nature* 427 (2004): 145–48.

168 **Thomas suggested that it would be useful:** Chris Thomas, "First Estimates of Extinction Risk from Climate Change," in *Saving a Million Species: Extinction Risk from Climate Change,* edited by Lee Jay Hannah (Washington, D.C.; Island Press, 2012), 17–18.

172 **perhaps as far as the mid-Miocene:** Aradhna K. Tripati, Christopher D. Roberts, and Robert E. Eagle, "Coupling of CO_2 and Ice Sheet Stability over Major Climate Transitions of the Last 20 Million Years," *Science* 326 (2009): 1394–97.

CHAPTER IX: ISLANDS ON DRY LAND

175 **"fragmentologist":** Jeff Tollefson, "Splinters of the Amazon," *Nature* 496 (2013): 286.

176 **"the most important ecological experiment":** Ibid.

176 **According to a recent study:** Roger LeB. Hooke, José F. Martín-Duque, and Javier Pedraza, "Land Transformation by Humans: A Review," *GSA Today* 22 (2012): 4–10.

176 **According to another recent study:** Erle C. Ellis and Navin Ramankutty, "Putting People in the Map: Anthropogenic Biomes of the World," *Frontiers in Ecology and the Environment* 6 (2008): 439–47.

179 **Across the reserves:** Richard O. Bierregard et al., *Lessons from Amazonia: The Ecology and Conservation of a Fragmented Forest* (New Haven, Conn.: Yale University Press, 2001), 41.

180 **On some land-bridge islands:** Jared Diamond, "The Island Dilemma: Lessons of Modern Biogeographic Studies for the Design of Natural Reserves," *Biological Conservation* 7 (1975): 129–46.

181 **the main predictor of local extinction:** Jared Diamond, "'Normal' Extinctions of Isolated Populations," in *Extinctions,* edited by Matthew H. Nitecki (Chicago: University of Chicago Press, 1984), 200.

181 **Researchers at the BDFFP have found:** Susan G. W. Laurance et al., "Effects of Road Clearings on Movement Patterns of Understory Rainforest Birds in Central Amazonia," *Conservation Biology* 18 (2004) 1099–109.

182 **"The jungle teems":** E. O. Wilson, *The Diversity of Life* (1992; reprint, New York: Norton, 1993), 3–4.

184 **There are butterflies that feed:** Carl W. Rettenmeyer et al., "The Largest Animal Association Centered on One Species: The Army Ant *Eciton burchellii* and Its More Than 300 Associates," *Insectes Sociaux* 58 (2011): 281–92.

184 **A pair of American naturalists:** Ibid.

185 **Erwin estimated:** Terry L. Erwin, "Tropical Forests: Their Richness in Coleoptera and Other Arthropod Species," *Coleopterists Bulletin* 36 (1982): 74–75.

185 **recent estimates suggest:** Andrew J. Hamilton et al., "Quantifying Uncertainty in Estimation of Tropical Arthropod Species Richness," *American Naturalist* 176 (2010): 90–95.

186 **This exact calculation:** E. O. Wilson, "Threats to Biodiversity," *Scientific American,* September 1989, 108–16.

186 **"Hardly a day passes":** John H. Lawton and Robert M. May, *Extinction Rates* (Oxford: Oxford University Press, 1995), v.

187 **A recent report by the Zoological Society of London:** "Spineless: Status and Trends of the World's Invertebrates," published online July 31, 2012, 17.

189 **"in the face of climatic change":** Thomas E. Lovejoy, "Biodiversity: What Is It?" in *Biodiversity II: Understanding and Protecting Our Biological Resources,* edited by Marjorie L. Kudla, Don E. Wilson, and E. O. Wilson (Washington, D.C.: Joseph Henry Press, 1997), 12.

CHAPTER X: THE NEW PANGAEA

196 **"the water I find":** Charles Darwin, letter to J. D. Hooker, Apr. 19, 1855, Darwin Correspondence Project, Cambridge University.

196 **But the results, he thought:** Charles Darwin, letter to *Gardeners' Chronicle,* May 21, 1855, Darwin Correspondence Project, Cambridge University.

197 **The tiny mollusks:** Darwin, *On the Origin of Species,* 385.

197 **It was no mere coincidence:** Ibid., 394.

197 **"The continents must have shifted":** Alfred Wegener, *The Origin of Continents and Oceans,* translated by John Biram (New York: Dover, 1966), 17.

198 **During any given twenty-four-hour period:** Mark A. Davis, *Invasion Biology* (Oxford: Oxford University Press, 2009), 22.

198 **"mass invasion event":** Anthony Ricciardi, "Are Modern Biological Invasions an Unprecedented Form of Global Change?" *Conservation Biology* 21 (2007): 329–36.

200 **The poet Randall Jarrell:** Randall Jarrell and Maurice Sendak, *The Bat-Poet* (1964; reprint, New York: HarperCollins, 1996), 1.

201 **It's also been proposed:** Paul M. Cryan et al., "Wing Pathology of White-Nose Syndrome in Bats Suggests Life-Threatening Disruption of Physiology," *BMC Biology* 8 (2010).

202 **In 1916:** This account of the Japanese beetle's spread comes from Charles S. Elton, *The Ecology of Invasions by Animals and Plants* (1958; reprint, Chicago: University of Chicago Press, 2000), 51–53.

202 **Roy van Driesche, an expert on invasive species:** Jason van Driesche and Roy van Driesche, *Nature out of Place: Biological Invasions in the Global Age* (Washington, D.C.: Island Press, 2000), 91.

203 **Of the more than seven hundred species:** Information on Hawaii's land snails comes from Christen Mitchell et al., *Hawaii's Comprehensive Wildlife*

Conservation Strategy (Honolulu: Department of Land and Natural Resources, 2005).

204 **"is precisely what *Homo sapiens*":** David Quammen, *The Song of the Dodo: Island Biogeography in an Age of Extinctions* (1996; reprint, New York: Scribner, 2004), 333.

204 **it made up close to half the standing timber:** Van Driesche and Van Driesche, *Nature out of Place*, 123.

204 **"not only was baby's crib":** George H. Hepting, "Death of the American Chestnut," *Forest and Conservation History* 18 (1974): 60.

206 **According to a study:** Paul Somers, "The Invasive Plant Problem," http://www.mass.gov/eea/docs/dfg/nhesp/land-protection-and-management/invasive-plant-problem.pdf.

206 **Although earthworms are beloved:** John C. Maerz, Victoria A. Nuzzo, and Bernd Blossey, "Declines in Woodland Salamander Abundance Associated with Non-Native Earthworm and Plant Invasions," *Conservation Biology* 23 (2009): 975–81.

207 **To dispose of the toads humanely:** "Operation Toad Day Out: Tip Sheet," Townsville City Council, <http://www.townsville.qld.gov.au/resident/pests/Documents/TDO%202012_Tip%20Sheet.pdf>.

207 **A recent study of visitors to Antarctica:** Steven L. Chown et al., "Continent-wide Risk Assessment for the Establishment of Nonindigenous Species in Antarctica," *Proceedings of the National Academy of Sciences* 109 (2012): 4938–43.

210 **"arguably the most successful invader":** Alan Burdick, *Out of Eden: An Odyssey of Ecological Invasion* (New York: Farrar, Straus and Giroux, 2005), 29.

210 **By the time humans pushed into North America:** Jennifer A. Leonard et al., "Ancient DNA Evidence for Old World Origin of New World Dogs," *Science* 298 (2002): 1613–16.

211 **"the introduction and acclimatization":** Quoted in Kim Todd, *Tinkering with Eden: A Natural History of Exotics in America* (New York: Norton, 2001), 137–38.

211 **According to the entry on pets:** Peter T. Jenkins, "Pet Trade," in *Encyclopedia of Biological Invasions*, edited by Daniel Simberloff and Marcel Rejmánek (Berkeley: University of California Press, 2011), 539–43.

211 **A recent study of non-indigenous species:** Gregory M. Ruiz et al., "Invasion of Coastal Marine Communities of North America: Apparent Patterns, Processes, and Biases," *Annual Review of Ecology and Systematics* 31 (2000): 481–531.

211 **For comparison's sake:** Van Driesche and Van Driesche, *Nature out of Place*, 46.

212 **"If we look far enough ahead":** Elton, *The Ecology of Invasions by Animals and Plants*, 50–51.

213 **In the case of terrestrial mammals:** James H. Brown, *Macroecology* (Chicago: University of Chicago Press, 1995), 220.

CHAPTER XI: THE RHINO GETS AN ULTRASOUND

218 **Genetic analysis has shown:** Ludovic Orlando et al., "Ancient DNA Analysis Reveals Woolly Rhino Evolutionary Relationships," *Molecular Phylogenetics and Evolution* 28 (2003): 485–99.

218 **a "living fossil":** E. O. Wilson, *The Future of Life* (2002; reprint, New York: Vintage, 2003), 80.

222 **Rhino horns . . . in recent years have become even more sought-after:** Adam Welz, "The Dirty War Against Africa's Remaining Rhinos," *e360*, published online Nov. 27, 2012.

223 **the population of African forest elephants:** Fiona Maisels et al., "Devastating Decline of Forest Elephants in Central Africa," *PLOS ONE* 8 (2013).

223 **as Tom Lovejoy has put it:** Thomas Lovejoy, "A Tsunami of Extinction," *Scientific American*, Dec. 2012, 33–34.

225 **"It shows us what the world might":** Tim F. Flannery, *The Future Eaters: An Ecological History of the Australasian Lands and People* (New York: G. Braziller, 1995), 55.

225 **the females were almost twice as giant:** Valérie A. Olson and Samuel T. Turvey, "The Evolution of Sexual Dimorphism in New Zealand Giant Moa (Dinornis) and Other Ratites," *Proceedings of the Royal Society B* 280 (2013).

226 **"We live in a zoologically impoverished world":** Alfred Russel Wallace, *The Geographical Distribution of Animals with a Study of the Relations of Living and Extinct Faunas as Elucidating the Past Changes of the Earth's Surface*, vol. 1 (New York: Harper and Brothers, 1876), 150.

227 **"At Big Bone Lick the first explorers":** Robert Morgan, "Big Bone Lick," posted online at: http://www.big-bone-lick.com/2011/10/.

228 **purchased for himself the teeth:** Charles Lyell, *Travels in North America, Canada, and Nova Scotia with Geological Observations*, 2nd ed. (London: J. Murray, 1855), 67.

229 **"great modification in climate":** Charles Lyell, *Geological Evidences of the Antiquity of Man, with Remarks on Theories of the Origin of Species by Variation*, 4th ed., revised (London: J. Murray, 1873), 189.

229 **"I cannot feel quite easy":** Quoted in Donald K. Grayson, "Nineteenth Century Explanations," in *Quaternary Extinctions: A Prehistoric Revolution*, edited by Paul S. Martin and Richard G. Klein (Tucson: University of Arizona Press, 1984), 32.

229 **"There must have been some physical cause":** Wallace, *The Geographical Distribution of Animals*, 150–51.

229 **"Looking at the whole subject again":** Alfred R. Wallace, *The World of Life: A Manifestation of Creative Power, Directive Mind and Ultimate Purpose* (New York: Moffat, Yard, 1911), 264.

231 **"When the chronology of extinction":** Paul S. Martin, "Prehistoric Overkill," in *Pleistocene Extinctions: The Search for a Cause,* edited by Paul S. Martin and H. E. Wright (New Haven, Conn.: Yale University Press, 1967), 115.

231 **"Personally, I can't fathom":** Jared Diamond, *Guns, Germs, and Steel: The Fates of Human Societies* (New York: Norton, 1997), 43.

231 **Then, rather abruptly:** Susan Rule et al., "The Aftermath of Megafaunal Extinction: Ecosystem Transformation in Pleistocene Australia," *Science* 335 (2012): 1483–86.

233 **Alroy has used computer simulations:** John Alroy, "A Multispecies Overkill Simulation of the End-Pleistocene Megafaunal Mass Extinction," *Science* 292 (2001): 1893–96.

234 **humans "are capable of driving":** John Alroy, "Putting North America's End-Pleistocene Megafaunal Extinction in Context," in *Extinctions in Near Time: Causes, Contexts, and Consequences,* edited by Ross D. E. MacPhee (New York: Kluwer Academic/Plenum, 1999), 138.

CHAPTER XII: THE MADNESS GENE

241 **"It must be admitted":** Charles Darwin, *The Descent of Man* (1871; reprint, New York: Penguin, 2004), 75.

241 **Boule invented:** James Shreeve, *The Neanderthal Enigma: Solving the Mystery of Human Origins* (New York: William Morrow, 1995), 38.

242 **"distinctly simian arrangement":** Marcellin Boule, *Fossil Men; Elements of Human Palaeontology,* translated by Jessie Elliot Ritchie and James Ritchie (Edinburgh: Oliver and Boyd, 1923), 224.

243 **would attract no more attention:** William L. Straus Jr. and A. J. E. Cave, "Pathology and the Posture of Neanderthal Man," *Quarterly Review of Biology* 32 (1957): 348–63.

243 **"We are brought suddenly to the realization":** Ray Solecki, *Shanidar, the First Flower People* (New York: Knopf, 1971), 250.

246 **In a paper published in *Science*:** Richard E. Green et al., "A Draft Sequence of the Neandertal Genome," *Science* 328 (2010): 710–22.

249 **When researchers from Leipzig:** E. Herrmann et al., "Humans Have Evolved Specialized Skills of Social Cognition: The Cultural Intelligence Hypothesis," *Science* 317 (2007): 1360–66.

253 **In a paper published . . . in *Nature*:** David Reich et al., "Genetic History of an Archaic Hominin Group from Denisova Cave in Siberia," *Nature* 468 (2010): 1053–60.

CHAPTER XIII: THE THING WITH FEATHERS

259 **"Futurology has never been":** Jonathan Schell, *The Fate of the Earth* (New York: Knopf, 1982), 21.

261 **"the problem of sharing our earth":** Carson, *Silent Spring*, 296.

266 **"human language is like the genetic code":** Michael Benton, "Paleontology and the History of Life," in *Evolution: The First Four Billion Years*, edited by Michael Ruse and Joseph Travis (Cambridge, Mass.: Belknap Press of Harvard University Press, 2009), 84.

268 **"*Homo sapiens* might not only be the agent":** Richard E. Leakey and Roger Lewin, *The Sixth Extinction: Patterns of Life and the Future of Humankind* (1995; reprint, New York: Anchor, 1996), 249.

268 **"Don't worry":** Annalee Newitz, *Scatter, Adapt, and Remember: How Humans Will Survive a Mass Extinction* (New York: Doubleday, 2013), 263.

SELECTED BIBLIOGRAPHY

Alroy, John. "A Multispecies Overkill Simulation of the End-Pleistocene Megafaunal Mass Extinction." *Science* 292 (2001): 1893–96.

Alvarez, Luis W. "Experimental Evidence That an Asteroid Impact Led to the Extinction of Many Species 65 Million Years Ago." *Proceedings of the National Academy of Sciences* 80 (1983): 627–42.

Alvarez, Luis W., W. Alvarez, F. Asaro, and H. V. Michel. "Extraterrestrial Cause for the Cretaceous-Tertiary Extinction." *Science* 208 (1980): 1095–108.

Alvarez, Walter. *T. rex and the Crater of Doom*. Princeton, N.J.: Princeton University Press, 1997.

———. "Earth History in the Broadest Possible Context." Ninety-Seventh Annual Faculty Research Lecture. University of California, Berkeley, International House, delivered Apr. 29, 2010.

Appel, Toby A. *The Cuvier-Geoffroy Debate: French Biology in the Decades Before Darwin*. New York: Oxford University Press, 1987.

Barnosky, Anthony D. "Megafauna Biomass Tradeoff as a Driver of Quaternary and Future Extinctions." *Proceedings of the National Academy of Sciences* 105 (2008): 11543–48.

———. *Heatstroke: Nature in an Age of Global Warming*. Washington, D.C.: Island Press/Shearwater Books, 2009.

Benton, Michael J. *When Life Nearly Died: The Greatest Mass Extinction of All Time*. New York: Thames and Hudson, 2003.

Bierregaard, Richard O., et al. *Lessons from Amazonia: The Ecology and Conservation of a Fragmented Forest*. New Haven, Conn.: Yale University Press, 2001.

Birkhead, Tim. "How Collectors Killed the Great Auk." *New Scientist* 142 (1994): 24–27.

Blundell, Derek J., and Andrew C. Scott, eds. *Lyell: The Past Is the Key to the Present.* London: Geological Society, 1998.

Bohor, B. F., et al. "Mineralogic Evidence for an Impact Event at the Cretaceous-Tertiary Boundary." *Science* 224 (1984): 867–69.

Boule, Marcellin. *Fossil Men: Elements of Human Palaeontology.* Translated by Jessie J. Elliot Ritchie and James Ritchie. Edinburgh: Oliver and Boyd, 1923.

Bowen, James, and Margarita Bowen. *The Great Barrier Reef: History, Science, Heritage.* Cambridge: Cambridge University Press, 2002.

Brown, James H. *Macroecology.* Chicago: University of Chicago Press, 1995.

Browne, Janet. *Charles Darwin: Voyaging.* New York: Knopf, 1995.

———. *Charles Darwin: The Power of Place.* New York: Knopf, 2002.

Browne, Malcolm W. "Dinosaur Experts Resist Meteor Extinction Idea." *New York Times*, Oct. 29, 1985.

Buckland, William. *Geology and Mineralogy Considered with Reference to Natural Theology.* London: W. Pickering, 1836.

Burdick, Alan. *Out of Eden: An Odyssey of Ecological Invasion.* New York: Farrar, Straus and Giroux, 2005.

Burkhardt, Richard Wellington. *The Spirit of System: Lamarck and Evolutionary Biology.* Cambridge, Mass.: Harvard University Press, 1977.

Butlin, Roger, Jon Bridle, and Dolph Schluter, eds. *Speciation and Patterns of Diversity.* Cambridge: Cambridge University Press, 2009.

Caldeira, Ken, and Michael E. Wickett. "Anthropogenic Carbon and Ocean pH." *Nature* 425 (2003): 365.

Carpenter, Kent E., et al. "One-Third of Reef-Building Corals Face Elevated Extinction Risk from Climate Change and Local Impacts." *Science* 321 (2008): 560–63.

Carson, Rachel. *Silent Spring.* 40th anniversary ed. Boston: Houghton Mifflin, 2002.

———. *The Sea Around Us.* Reprint, New York: Signet, 1961.

Catenazzi, Alessandro, et al. "*Batrachochytrium dendrobatidis* and the Collapse of Anuran Species Richness and Abundance in the Upper Manú National Park, Southeastern Peru." *Conservation Biology* 25 (2011) 382–91.

Chown, Steven L., et al. "Continent-wide Risk Assessment for the Establishment of Nonindigenous Species in Antarctica." *Proceedings of the National Academy of Sciences* 109 (2012): 4938–43.

Chu, Jennifer. "Timeline of a Mass Extinction." MIT News Office, published online Nov. 18, 2011.

Cohen, Claudine. *The Fate of the Mammoth: Fossils, Myth, and History.* Chicago: University of Chicago Press, 2002.

Coleman, William. *Georges Cuvier, Zoologist: A Study in the History of Evolution Theory*. Cambridge, Mass.: Harvard University Press, 1964.

Collen, Ben, Monika Böhm, Rachael Kemp, and Jonathan E. M. Baillie, eds. *Spineless: Status and Trends of the World's Invertebrates*. London: Zoological Society, 2012.

Collinge, Sharon K. *Ecology of Fragmented Landscapes*. Baltimore: Johns Hopkins University Press, 2009.

Collins, James P., and Martha L. Crump. *Extinctions in Our Times: Global Amphibian Decline*. Oxford: Oxford University Press, 2009.

Crump, Martha L. *In Search of the Golden Frog*. Chicago: University of Chicago Press, 2000.

Crutzen, Paul J. "Geology of Mankind." *Nature* 415 (2002): 23.

Cryan, Paul M., et al. "Wing Pathology of White-Nose Syndrome in Bats Suggests Life-Threatening Disruption of Physiology." *BMC Biology* 8 (2010).

Cuvier, Georges, and Martin J. S. Rudwick. *Georges Cuvier, Fossil Bones, and Geological Catastrophes: New Translations and Interpretations of the Primary Texts*. Chicago: University of Chicago Press, 1997.

Darwin, Charles. *The Structure and Distribution of Coral Reefs*. 3rd ed. New York: D. Appleton, 1897.

———. *On the Origin of Species: A Facsimile of the First Edition*. Cambridge, Mass.: Harvard University Press, 1964.

———. *The Autobiography of Charles Darwin, 1809–1882: With Original Omissions Restored*. New York: Norton, 1969.

———. *The Works of Charles Darwin*. Vol. 1, *Diary of the Voyage of H.M.S.* Beagle. Edited by Paul H. Barrett and R. B. Freeman. New York: New York University Press, 1987

———. *The Works of Charles Darwin*. Vol. 2, *Journal of Researches*. Edited by Paul H. Barrett and R. B. Freeman. New York: New York University Press, 1987.

———. *The Works of Charles Darwin*. Vol. 3, *Journal of Researches*, Part 2. Edited by Paul H. Barrett and R. B. Freeman. New York: New York University Press, 1987.

———. *The Descent of Man*. 1871. Reprint, New York: Penguin, 2004.

Davis, Mark A. *Invasion Biology*. Oxford: Oxford University Press, 2009.

De'ath, Glenn, et al. "The 27-Year Decline of Coral Cover on the Great Barrier Reef and Its Causes." *Proceedings of the National Academy of Sciences* 109 (2012): 17995–99.

DeWolf, Gordon P. *Native and Naturalized Trees of Massachusetts*. Amherst: Cooperative Extension Service, University of Massachusetts, 1978.

Diamond, Jared. "The Island Dilemma: Lessons of Modern Biogeographic Studies for the Design of Natural Reserves." *Biological Conservation* 7 (1975): 129–46.

Diamond, Jared. *Guns, Germs, and Steel: The Fates of Human Societies*. New York: Norton, 2005.

Dobbs, David. *Reef Madness: Charles Darwin, Alexander Agassiz, and the Meaning of Coral*. New York: Pantheon, 2005.

Ellis, Erle C., and Navin Ramankutty. "Putting People in the Map: Anthropogenic Biomes of the World." *Frontiers in Ecology and the Environment* 6 (2008): 439–47.

Elton, Charles S. *The Ecology of Invasions by Animals and Plants*. 1958. Reprint, Chicago: University of Chicago Press, 2000.

Erwin, Douglas H. *Extinction: How Life on Earth Nearly Ended 250 Million Years Ago*. Princeton, N.J.: Princeton University Press, 2006.

Erwin, Terry L. "Tropical Forests: Their Richness in Coleoptera and Other Arthropod Species." *Coleopterists Bulletin* 36 (1982): 74–75.

Fabricius, Katherina E., et al. "Losers and Winners in Coral Reefs Acclimatized to Elevated Carbon Dioxide Concentrations." *Nature Climate Change* 1 (2011): 165–69.

Feeley, Kenneth J., et al. "Upslope Migration of Andean Trees." *Journal of Biogeography* 38 (2011): 783–91.

Feeley, Kenneth J., and Miles R. Silman. "Biotic Attrition from Tropical Forests Correcting for Truncated Temperature Niches." *Global Change Biology* 16 (2010): 1830–36.

Flannery, Tim F. *The Future Eaters: An Ecological History of the Australasian Lands and People*. New York: G. Braziller, 1995.

Fortey, Richard A. *Life: A Natural History of the First Four Billion Years of Life on Earth*. New York: Vintage, 1999.

Fuller, Errol. *The Great Auk*. New York: Abrams, 1999.

Gaskell, Jeremy. *Who Killed the Great Auk?* Oxford: Oxford University Press, 2000.

Gattuso, Jean-Pierre, and Lina Hansson, eds. *Ocean Acidification*. Oxford: Oxford University Press, 2011.

Gleick, James. *Chaos: Making a New Science*. New York: Viking, 1987.

Glen, William, ed. *The Mass-Extinction Debates: How Science Works in a Crisis*. Stanford, Calif.: Stanford University Press, 1994.

Goodell, Jeff. *How to Cool the Planet: Geoengineering and the Audacious Quest to Fix Earth's Climate*. Boston: Houghton Mifflin Harcourt, 2010.

Gould, Stephen Jay. *The Panda's Thumb: More Reflections in Natural History*. New York: Norton, 1980.

Grant, K. Thalia, and Gregory B. Estes. *Darwin in Galápagos: Footsteps to a New World*. Princeton, N.J.: Princeton University Press, 2009.

Grayson, Donald K., and David J. Meltzer. "A Requiem for North American Overkill." *Journal of Archaeological Science* 30 (2003): 585–93.

Green, Richard E., et al. "A Draft Sequence of the Neandertal Genome." *Science* 328 (2010): 710–22.

Hallam, A. *Great Geological Controversies.* Oxford: Oxford University Press, 1983.

Hallam, A., and P. B. Wignall. *Mass Extinctions and Their Aftermath.* Oxford: Oxford University Press, 1997.

Hall-Spencer, Jason M., et al. "Volcanic Carbon Dioxide Vents Show Ecosystem Effects of Ocean Acidification." *Nature* 454 (2008): 96–99.

Hamilton, Andrew J., et al. "Quantifying Uncertainty in Estimation of Tropical Arthropod Species Richness." *American Naturalist* 176 (2010): 90–95.

Hannah, Lee Jay, ed. *Saving a Million Species: Extinction Risk from Climate Change.* Washington, D.C.: Island Press, 2012.

Haynes, Gary, ed. *American Megafaunal Extinctions at the End of the Pleistocene.* Dordrecht: Springer, 2009.

Heatwole, Harold, Terence Done, and Elizabeth Cameron. *Community Ecology of a Coral Cay: A Study of One Tree Island, Great Barrier Reef, Australia.* The Hague: W. Junk, 1981.

Hedeen, Stanley. *Big Bone Lick: The Cradle of American Paleontology.* Lexington: University Press of Kentucky, 2008.

Hepting, George H. "Death of the American Chestnut." *Forest and Conservation History* 18 (1974): 60–67.

Herbert, Sandra. *Charles Darwin, Geologist.* Ithaca, N.Y.: Cornell University Press, 2005.

Herrmann, E., et al. "Humans Have Evolved Specialized Skills of Social Cognition: The Cultural Intelligence Hypothesis." *Science* 317 (2007): 1360–66.

Hoegh-Guldberg, Ove, et al. "Coral Reefs under Rapid Climate Change and Ocean Acidification." *Science* 318 (2007): 1737–42.

Hoffmann, Michael, et al. "The Impact of Conservation on the Status of the World's Vertebrates." *Science* 330 (2010): 1503–9.

Holdaway, Richard N., and Christopher Jacomb. "Rapid Extinction of the Moas (Aves: Dinornithiformes): Model, Test, and Implications." *Science* 287 (2000): 2250–54.

Hooke, Roger, José F. Martín-Duque, and Javier Pedraza. "Land Transformation by Humans: A Review." *GSA Today* 22 (2012): 4–10.

Huggett, Richard J. *Catastrophism: Systems of Earth History.* London: E. Arnold, 1990.

Humboldt, Alexander von. *Views of Nature, or, Contemplations on the Sublime Phenomena of Creation with Scientific Illustrations.* Translated by Elsie C. Otté and Henry George Bohn. London: H. G. Bohn, 1850.

Humboldt, Alexander von, and Aimé Bonpland. *Essay on the Geography of Plants.* Edited by Stephen T. Jackson. Translated by Sylvie Romanowski. Chicago: University of Chicago Press, 2008.

Hunt, Terry L. "Rethinking Easter Island's Ecological Catastrophe." *Journal of Archaeological Science* 34 (2007): 485–502.

Hutchings, P. A., Michael Kingsford, and Ove Hoegh-Guldberg, eds. *The Great Barrier Reef: Biology, Environment and Management*. Collingwood, Australia: CSIRO, 2008.

Janzen, Daniel H. "Why Mountain Passes Are Higher in the Tropics." *American Naturalist* 101 (1967): 233–49.

Jarrell, Randall, and Maurice Sendak. *The Bat-Poet*. 1964. Reprint, New York: HarperCollins, 1996.

Johnson, Chris. *Australia's Mammal Extinctions: A 50,000 Year History*. Cambridge: Cambridge University Press, 2006.

Kiessling, Wolfgang, and Carl Simpson. "On the Potential for Ocean Acidification to Be a General Cause of Ancient Reef Crises." *Global Change Biology* 17 (2011): 56–67.

Knoll, A. H. "Biomineralization and Evolutionary History." *Reviews in Mineralogy and Geochemistry* 54 (2003): 329–56.

Kudla, Marjorie L., Don E. Wilson, and E. O. Wilson, eds. *Biodiversity II: Understanding and Protecting Our Biological Resources*. Washington, D.C.: Joseph Henry Press, 1997.

Kuhn, Thomas S. *The Structure of Scientific Revolutions*. 2nd ed. Chicago: University of Chicago Press, 1970.

Kump, Lee, Timothy Bralower, and Andy Ridgwell. "Ocean Acidification in Deep Time." *Oceanography* 22 (2009): 94–107.

Kump, Lee R., Alexander Pavlov, and Michael A. Arthur. "Massive Release of Hydrogen Sulfide to the Surface Ocean and Atmosphere during Intervals of Oceanic Anoxia." *Geology* 33 (2005): 397.

Landman, Neil, et al. "Mode of Life and Habitat of Scaphitid Ammonites." *Geobios* 54 (2012): 87–98.

Laurance, Susan G. W., et al. "Effects of Road Clearings on Movement Patterns of Understory Rainforest Birds in Central Amazonia." *Conservation Biology* 18 (2004): 1099–109.

Lawton, John H., and Robert M. May. *Extinction Rates*. Oxford: Oxford University Press, 1995.

Leakey, Richard E., and Roger Lewin. *The Sixth Extinction: Patterns of Life and the Future of Humankind*. 1995. Reprint, New York: Anchor, 1996.

Lee, R. *Memoirs of Baron Cuvier*. New York: J. and J. Harper, 1833.

Lenton, Timothy M., et al. "First Plants Cooled the Ordovician." *Nature Geoscience* 5 (2012): 86–9.

Levy, Sharon. *Once and Future Giants: What Ice Age Extinctions Tell Us about the Fate of Earth's Largest Animals*. Oxford: Oxford University Press, 2011.

Longrich, Nicholas R., Bhart-Anjan S. Bhullar, and Jacques A. Gauthier. "Mass

Extinction of Lizards and Snakes at the Cretaceous-Paleogene Boundary." *Proceedings of the National Academy of Sciences* 109 (2012): 21396–401.

Longrich, Nicholas R., T. Tokaryk, and D. J. Field. "Mass Extinction of Birds at the Cretaceous-Paleogene (K-Pg) Boundary." *Proceedings of the National Academy of Sciences* 108 (2011): 15253–57.

Lopez, Barry. *Arctic Dreams*. 1986. Reprint, New York: Vintage, 2001.

Lovejoy, Thomas. "A Tsunami of Extinction." *Scientific American*, Dec. 2012, 33–34.

Lyell, Charles. *Travels in North America, Canada, and Nova Scotia with Geological Observations*. 2nd ed. London: J. Murray, 1855.

———. *Geological Evidences of the Antiquity of Man; with Remarks on Theories of the Origin of Species by Variation*. 4th ed, revised. London: Murray, 1873.

———. *Life, Letters and Journals of Sir Charles Lyell*, edited by Mrs. Lyell. London: J. Murray, 1881.

———. *Principles of Geology*. Vol. 1. Chicago: University of Chicago Press, 1990.

———. *Principles of Geology*. Vol. 2. Chicago: University of Chicago Press, 1990.

———. *Principles of Geology*. Vol. 3. Chicago: University of Chicago Press, 1991.

MacPhee, R. D. E., ed. *Extinctions in Near Time: Causes, Contexts, and Consequences*. New York: Kluwer Academic/Plenum, 1999.

Maerz, John C., Victoria A. Nuzzo, and Bernd Blossey. "Declines in Woodland Salamander Abundance Associated with Non-Native Earthworm and Plant Invasions." *Conservation Biology* 23 (2009): 975–81.

Maisels, Fiona, et al. "Devastating Decline of Forest Elephants in Central Africa." *PLOS ONE* 8 (2013).

Martin, Paul S., and Richard G. Klein, eds. *Quaternary Extinctions: A Prehistoric Revolution*. Tucson: University of Arizona Press, 1984.

Martin, Paul S., and H. E. Wright, eds. *Pleistocene Extinctions: The Search for a Cause*. New Haven, Conn.: Yale University Press, 1967.

Marvin, Ursula B. *Continental Drift: The Evolution of a Concept*. Washington, D.C.: Smithsonian Institution Press (distributed by G. Braziller), 1973.

Mayr, Ernst. *The Growth of Biological Thought: Diversity, Evolution, and Inheritance*. Cambridge, Mass.: Belknap Press of Harvard University Press, 1982.

McCallum, Malcolm L. "Amphibian Decline or Extinction? Current Declines Dwarf Background Extinction Rates." *Journal of Herpetology* 41 (2007): 483–91.

McKibben, Bill. *The End of Nature*. New York: Random House, 1989.

Mendelson, Joseph R. "Shifted Baselines, Forensic Taxonomy, and Rabb's Fringe-limbed Treefrog: The Changing Role of Biologists in an Era of Amphibian Declines and Extinctions." *Herpetological Review* 42 (2011): 21–25.

Mitchell, Alanna. *Seasick: Ocean Change and the Extinction of Life on Earth*. Chicago: University of Chicago Press, 2009.

Mitchell, Christen, et al. *Hawaii's Comprehensive Wildlife Conservation Strategy*. Honolulu: Department of Land and Natural Resources, 2005.

Mittelbach, Gary G., et al. "Evolution and the Latitudinal Diversity Gradient: Speciation, Extinction and Biogeography." *Ecology Letters* 10 (2007): 315–31.

Monks, Neale, and Philip Palmer. *Ammonites*. Washington, D.C.: Smithsonian Institution Press, 2002.

Moum, Truls, et al. "Mitochondrial DNA Sequence Evolution and Phylogeny of the Atlantic Alcidae, Including the Extinct Great Auk (*Pinguinus impennis*)." *Molecular Biology and Evolution* 19 (2002): 1434–39.

Muller, Richard. *Nemesis*. New York: Weidenfeld and Nicolson, 1988.

Musgrave, Ruth A. "Incredible Frog Hotel." *National Geographic Kids*, Sept. 2008, 16–19.

Newitz, Annalee. *Scatter, Adapt, and Remember: How Humans Will Survive a Mass Extinction*. New York: Doubleday, 2013.

Newman, M. E. J., and Richard G. Palmer. *Modeling Extinction*. Oxford: Oxford University Press, 2003.

Newton, Alfred. "Abstract of Mr. J. Wolley's Researches in Iceland Respecting the Gare-Fowl or Great Auk." *Ibis* 3 (1861): 374–99.

Nitecki, Matthew H., ed. *Extinctions*. Chicago: University of Chicago Press, 1984.

Novacek, Michael J. *Terra: Our 100-Million-Year-Old Ecosystem—and the Threats That Now Put It at Risk*. New York: Farrar, Straus and Giroux, 2007.

Olson, Valérie A., and Samuel T. Turvey. "The Evolution of Sexual Dimorphism in New Zealand Giant Moa (Dinornis) and Other Ratites." *Proceedings of the Royal Society B* 280 (2013).

Orlando, Ludovic, et al. "Ancient DNA Analysis Reveals Woolly Rhino Evolutionary Relationships." *Molecular Phylogenetics and Evolution* 28 (2003): 485–99.

Outram, Dorinda. *Georges Cuvier: Vocation, Science and Authority in Post-Revolutionary France*. Manchester, England: Manchester University Press, 1984.

Palmer, Trevor. *Perilous Planet Earth: Catastrophes and Catastrophism through the Ages*. Cambridge: Cambridge University Press, 2003.

Peale, Charles Willson. *The Selected Papers of Charles Willson Peale and His Family*. Edited by Lillian B. Miller, Sidney Hart, and Toby A. Appel. New Haven, Conn.: Yale University Press (published for the National Portrait Gallery, Smithsonian Institution), 1983–2000.

Phillips, John. *Life on the Earth*. Cambridge: Macmillan and Company, 1860.

Plaisance, Laetitia, et al. "The Diversity of Coral Reefs: What Are We Missing?" *PLOS ONE* 6 (2011).

Powell, James Lawrence. *Night Comes to the Cretaceous: Dinosaur Extinction and the Transformation of Modern Geology*. New York: W. H. Freeman, 1998.

Quammen, David. *The Song of the Dodo: Island Biogeography in an Age of Extinctions*. 1996. Reprint, New York: Scribner, 2004.

———. *The Reluctant Mr. Darwin: An Intimate Portrait of Charles Darwin and the Making of His Theory of Evolution*. New York: Atlas Books/Norton, 2006.

———. *Natural Acts: A Sidelong View of Science and Nature*. Revised ed., New York: Norton, 2008.

Rabinowitz, Alan. "Helping a Species Go Extinct: The Sumatran Rhino in Borneo." *Conservation Biology* 9 (1995): 482–88.

Randall, John E., Gerald R. Allen, and Roger C. Steene. *Fishes of the Great Barrier Reef and Coral Sea*. Honolulu: University of Hawaii Press, 1990.

Raup, David M. *The Nemesis Affair: A Story of the Death of Dinosaurs and the Ways of Science*. New York: Norton, 1986.

———. *Extinction: Bad Genes or Bad Luck?* New York: Norton, 1991.

Raup, David M., and J. John Sepkoski Jr. "Periodicity of Extinctions in the Geologic Past." *Proceedings of the National Academy of Sciences* 81 (1984): 801–5.

———. "Mass Extinctions in the Marine Fossil Record." *Science* 215 (1982): 1501–3.

Reich, David, et al. "Genetic History of an Archaic Hominin Group from Denisova Cave in Siberia." *Nature* 468 (2010): 1053–60.

Rettenmeyer, Carl W. et al. "The Largest Animal Association Centered on One Species: The Army Ant *Eciton burchellii* and Its More Than 300 Associates." *Insectes Sociaux* 58 (2011): 281–92.

Rhodes, Frank H. T., Richard O. Stone, and Bruce D. Malamud. *Language of the Earth: A Literary Anthology*. 2nd ed. Chichester, England: Wiley, 2009.

Ricciardi, Anthony. "Are Modern Biological Invasions an Unprecedented Form of Global Change?" *Conservation Biology* 21 (2007): 329–36.

Rose, Kenneth D. *The Beginning of the Age of Mammals*. Baltimore: Johns Hopkins University Press, 2006.

Rosenzweig, Michael L. *Species Diversity in Space and Time*. Cambridge: Cambridge University Press, 1995.

Rudwick, M. J. S. *The Meaning of Fossils: Episodes in the History of Palaeontology*. 2nd revised ed. New York: Science History, 1976.

———. *Bursting the Limits of Time: The Reconstruction of Geohistory in the Age of Revolution*. Chicago: University of Chicago Press, 2005.

———. *Lyell and Darwin, Geologists: Studies in the Earth Sciences in the Age of Reform*. Aldershot, England: Ashgate, 2005.

———. *Worlds Before Adam: The Reconstruction of Geohistory in the Age of Reform*. Chicago: University of Chicago Press, 2008.

Ruiz, Gregory M., et al. "Invasion of Coastal Marine Communities in North America: Apparent Patterns, Processes, and Biases." *Annual Review of Ecology and Systematics* 31 (2000): 481–531.

Rule, Susan, et al. "The Aftermath of Megafaunal Extinction: Ecosystem Transformation in Pleistocene Australia." *Science* 335 (2012): 1483–86.

Ruse, Michael, and Joseph Travis, eds. *Evolution: The First Four Billion Years*. Cambridge, Mass.: Belknap Press of Harvard University Press, 2009.

Schell, Jonathan. *The Fate of the Earth*. New York: Knopf, 1982.

Sellers, Charles Coleman. *Mr. Peale's Museum: Charles Willson Peale and the First Popular Museum of Natural Science and Art*. New York: Norton, 1980.

Semonin, Paul. *American Monster: How the Nation's First Prehistoric Creature Became a Symbol of National Identity*. New York: New York University Press, 2000.

Severance, Frank H. *An Old Frontier of France: The Niagara Region and Adjacent Lakes under French Control*. New York: Dodd, 1917.

Shen, Shu-zhong, et al. "Calibrating the End-Permian Mass Extinction." *Science* 334 (2011): 1367–72.

Sheppard, Charles, Simon K. Davy, and Graham M. Pilling. *The Biology of Coral Reefs*. Oxford: Oxford University Press, 2009.

Shreeve, James. *The Neandertal Enigma: Solving the Mystery of Modern Human Origins*. New York: William Morrow, 1995.

Shrenk, Friedemann, and Stephanie Müller. *The Neanderthals*. London: Routledge, 2009.

Silverman, Jacob, et al. "Coral Reefs May Start Dissolving when Atmospheric CO_2 Doubles." *Geophysical Research Letters* 35 (2009).

Simberloff, Daniel, and Marcel Rejmánek, eds., *Encyclopedia of Biological Invasions*. Berkeley: University of California Press, 2011.

Simpson, George Gaylord. *Why and How: Some Problems and Methods in Historical Biology*. Oxford: Pergamon Press, 1980.

Soto-Azat, Claudio, et al. "The Population Decline and Extinction of Darwin's Frogs." *PLOS ONE* 8 (2013).

Stanley, Steven M. *Extinction*. New York: Scientific American Library, 1987.

Stolzenburg, William. *Rat Island: Predators in Paradise and the World's Greatest Wildlife Rescue*. New York: Bloomsbury, 2011.

Straus, William L., Jr., and A. J. E. Cave. "Pathology and the Posture of Neanderthal Man." *Quarterly Review of Biology* 32 (1957): 348–63.

Sulloway, Frank J. "Darwin and His Finches: The Evolution of a Legend." *Journal of the History of Biology* 15 (1982): 1–53.

Taylor, Paul D. *Extinctions in the History of Life*. Cambridge: Cambridge University Press, 2004.

Thomas, Chris D., et al. "Extinction Risk from Climate Change." *Nature* 427 (2004): 145–48.

Thomson, Keith Stewart. *The Legacy of the Mastodon: The Golden Age of Fossils in America*. New Haven, Conn.: Yale University Press, 2008.

Todd, Kim. *Tinkering with Eden: A Natural History of Exotics in America*. New York: Norton, 2001.

Tollefson, Jeff. "Splinters of the Amazon." *Nature* 496 (2013): 286–89.

Tripati, Aradhna K., Christopher D. Roberts, and Robert A. Eagle. "Coupling of

CO_2 and Ice Sheet Stability over Major Climate Transitions of the Last 20 Million Years." *Science* 326 (2009): 1394–97.

Turvey, Samuel. *Holocene Extinctions*. Oxford: Oxford University Press, 2009.

Urrutia, Rocío, and Mathias Vuille. "Climate Change Projections for the Tropical Andes Using a Regional Climate Model: Temperature and Precipitation Simulations for the End of the 21st Century." *Journal of Geophysical Research* 114 (2009).

Van Driesche, Jason, and Roy Van Driesche. *Nature out of Place: Biological Invasions in the Global Age*. Washington, D.C.: Island Press, 2000.

Veron, J. E. N. *A Reef in Time: The Great Barrier Reef from Beginning to End*. Cambridge, Mass.: Belknap Press of Harvard University Press, 2008.

———. "Is the End in Sight for the World's Coral Reefs?" *e360*, published online Dec. 6, 2010.

Wake, D. B., and V. T. Vredenburg. "Colloquium Paper: Are We in the Midst of the Sixth Mass Extinction? A View from the World of Amphibians." *Proceedings of the National Academy of Sciences* 105 (2008): 11466–73.

Wallace, Alfred Russel. *The Geographical Distribution of Animals with a Study of the Relations of Living and Extinct Faunas as Elucidating the Past Changes of the Earth's Surface*. Vol. 1. New York: Harper and Brothers, 1876.

———. *Tropical Nature and Other Essays*. London: Macmillan, 1878.

———. *The Wonderful Century: Its Successes and Its Failures*. New York: Dodd, Mead, 1898.

———. *The World of Life: A Manifestation of Creative Power, Directive Mind and Ultimate Purpose*. New York: Moffat, Yard, 1911.

Wegener, Alfred. *The Origin of Continents and Oceans*. Translated by John Biram. New York: Dover, 1966.

Wells, Kentwood David. *The Ecology and Behavior of Amphibians*. Chicago: University of Chicago Press, 2007.

Welz, Adam. "The Dirty War against Africa's Remaining Rhinos." *e360*, published online Nov. 27, 2012.

Whitfield, John. *In the Beat of a Heart: Life, Energy, and the Unity of Nature*. Washington, D.C.: National Academies Press, 2006.

Whitmore, T. C., and Jeffrey Sayer, eds. *Tropical Deforestation and Species Extinction*. London: Chapman and Hall, 1992.

Wilson, Edward O. "Threats to Biodiversity." *Scientific American*, Sept. 1989, 108–16.

———. *The Diversity of Life*. 1992. Reprint, New York: Norton, 1993.

———. *The Future of Life*. 2002. Reprint, New York: Vintage, 2003.

Wilson, Leonard G. *Charles Lyell, the Years to 1841: The Revolution in Geology*. New Haven, Conn.: Yale University Press, 1972.

Wollaston, Alexander F. R. *Life of Alfred Newton*. New York: E. P. Dutton, 1921.

Worthy, T. H., and Richard N. Holdaway. *The Lost World of the Moa: Prehistoric Life of New Zealand*. Bloomington: Indiana University Press, 2002.

Zalasiewicz, Jan. *The Earth After Us: What Legacy Will Humans Leave in the Rocks?* Oxford: Oxford University Press, 2008.

Zalasiewicz, Jan, et al. "Are We Now Living in the Anthropocene?" *GSA Today* 18 (2008): 4–8.

Zalasiewicz, Jan, et al. "Graptolites in British Stratigraphy." *Geological Magazine* 146 (2009): 785–850.

PHOTO/ILLUSTRATION CREDITS

7 © Vance Vredenburg
9 © Michael & Patricia Fogden/Minden Pictures
16 adapted from David M. Raup and J. John Sepkoski Jr./*Science* 215 (1982), 1502
32 Paul D. Stewart/Science Source
37 Reproduction by permission of the Rare Book Room, Buffalo and Erie County Public Library, Buffalo, New York
39 The Granger Collection, New York
40 © The British Library Board, 39.i.15 pl.1
49 Hulton Archive/Getty Images
61 © Natural History Museum, London/Mary Evans Picture Library
64 Matthew Kleiner
66 Natural History Museum/Science Source
71 Elizabeth Kolbert
73 © ER Degginger/Science Source
77 adapted from John Phillips's *Life on Earth*
81 Detlev van Ravenswaay/Science Source
84 NASA/GSFC/DLR/ASU/Science Source
89 Used with permission from the Paleontological Society
95 John Scott/E+/Getty Images

98 British and Irish Graptolite Group
100 EMR Wood/Palaeontolographical Society
115 John Kleiner
119 Steve Gschmeissner/Science Source
126 © Gary Bell/OceanwideImages.com
129 Nancy Sefton/Science Source
146 David Doubilet/National Geographic/Getty Images
150 © Miles R. Silman
155 © William Farfan Rios
157 © Miles Silman
175 The Biological Dynamics of Forest Fragments Project
184 © Konrad Wothe/Minden Pictures
191 © Philip C. Stouffer
195 U.S. Fish and Wildlife Service/Science Source
207 John Kleiner
215 Vermont Fish and Wildlife Department/Joel Flewelling
221 Tom Uhlman, Cincinnati Zoo
226 © Natural History Museum, London/Mary Evans Picture Library
230 AFP/Getty Images/Newscom
238 Neanderthal Museum
242 © Paul D. Stewart/NPL/Minden Pictures
244 Neanderthal Museum
252 Thanks to Ed Green
264 San Diego Zoo

Photo research by Laura Wyss and Wyssphoto, Inc.
Images on pages 16, 77, 89, 165, 271 rendered by Mapping Specialists

INDEX

Page numbers in *italics* refer to illustrations.

ABOUT THE AUTHOR

ELIZABETH KOLBERT is a staff writer at the *New Yorker.* She is the author of *Field Notes from a Catastrophe: Man, Nature, and Climate Change.* She lives in Williamstown, Massachusetts, with her husband and children.